READING AND WRITING SKILLS IN PRIMARY EDUCATION

COUNCIL OF EUROPE

READING AND WRITING SKILLS IN PRIMARY EDUCATION

A REPORT OF THE EDUCATIONAL RESEARCH
WORKSHOP HELD IN TILBURG (THE NETHERLANDS)
9-12 DECEMBER 1986

EDITED BY
MAY YOUNG, MARIE THOMAS AND PAMELA MUNN
OF THE SCOTTISH COUNCIL FOR RESEARCH IN EDUCATION
IN CO-OPERATION WITH
PROF. DR. L.F.W. DE KLERK, TILBURG UNIVERSITY

Taylor & Francis
Taylor & Francis Group

LONDON AND NEW YORK

Library of Congress Cataloging-in-Publication Data

Educational Research Workshop (1986 : Tilburg, Netherlands)
Reading and writing skills in primary education : a report of the
Educational Research Workshop held in Tilburg (the Netherlands) 9-12
December 1986 / edited by the Scottish Council for Research in
Education in co-operation with L.F.W. de Klerk.
    p. cm.
At head of title: Council of Europe.
Includes bibliographies.
ISBN 9026508808 (Taylor & Francis)
1. Language arts (Primary)--Europe--Congresses. 2. Reading
(Primary)--Europe--Congresses. I. Klerk, L.F.W. de.
II. Scottish Council for Research in Education. III. Council of
Europe. IV. Title.
LB1529.E85E38 1986
372.6'094--dc19   87-18171
                                                                 CIP

Published by Taylor & Francis
2 Park Square, Milton Park, Abingdon, Oxon, OX14 4RN
270 Madison Ave, New York NY 10016

Transferred to Digital Printing 2007

CIP-gegevens Koninklijke Bibliotheek, Den Haag

Reading

Reading and writing skills in primary education : a report
of the educational research workshop held in Tilburg (The
Netherlands) 9-12 December 1986 / ed. by the Scottish
Council for Research in Education in co-operation with
L.F.W. de Klerk. – Berwyn : Swets North America ; Lisse :
Taylor & Francis
Publ. by the Council for Cultural Co-operation (CDCC) of
the Council of Europe. – Met lit. opg.
ISBN 90-265-0880-8
SISO 475.1 UDC 372.41/.45(4) NUGI 722
Trefw.: leesonderwijs ; Europa / schrijfonderwijs ; Europa.

372.6' O94--dc19

ISBN 90 265 0880 8
NUGI 722

Printed and bound by CPI Antony Rowe, Eastbourne

Publisher's Note
The publisher has gone to great lengths to ensure the quality of this reprint
but points out that some imperfections in the original may be apparent

CONTENTS

PREFACE

The Tilburg Workshop was one of a series of educational research meetings which have become an inportant element in the programme of the Council for Cultural Co-operation of the Council of Europe since 1975. European co-operation in educational research aims at providing Ministries of Education with research findings so as to enable them to prepare their policy decisions. Co-operation should also lead to a joint European evaluation of certain educational reforms. The educational research meetings bring together resear-chers from the 24 countries taking part in the work of the Council for Cultural Co-operation. The purpose is to compare research findings on a particular topic of current interest; to identify areas of research so far neglected and to discuss possibilities for joint research projects. The reports, as well as a selection of the papers of these meetings, are usually published as a book so that Ministries and research workers, as well as a wider public (teachers, parents, press) are kept informed of the present state of research at European level.

The meeting in Tilburg goes back to a suggestion made by the Project Group, "Innovation in primary education," subsequently taken up by the CDCC. The Institute for Educational Research in the Netherlands (het Instituut voor Onderzoek van het Onderwijs, SVO) kindly agreed to organise the Workshop, in co-operation with the CDCC, and the University of Tilburg offered to host it.

The theme "Writing and reading skills in primary education" was chosen because of the growing functional illiteracy among school leavers in Europe. Although schools alone may not be able to cope with this problem, it was felt that a new approach to reading and writing in primary education might help to predict reading difficulties and prevent illiteracy through special support for poor readers.

The Workshop took place in the Senate Room of the University. Seven commissioned papers (covering Belgium, Denmark, France, the Federal Repu-blic of Germany, the Netherlands, Sweden and the United Kingdom) were presented in plenary session and then discussed in three working groups. National and individual reports from a number of countries, as well as lists of research projects and bibliographines, were tabled as background material. On the final day the Rapporteur General, Professor Dr. L.F.W. de Klerk from Tilburg University, summed up the situation and the conclusions as he interpreted them.

The following countries were represented: Austria, Belgium, Cyprus, Finland, France, the Federal Republic of Germany, Ireland, Malta, the Netherlands, Portugal, Spain, Sweden, Switzerland and the United Kingdom. There were also observers from Hungary, Yugoslavia and WCOTP. The list of participants is given at the end of this book.

The Council of Europe is particularly grateful to the Institute for Educational Research in the Netherlands (Dr. J.G.L.C. Lodewijks and Drs. C.M.R. Verkoeijen), as well as to the local organisers in Tilburg (Dr. R.J. Simons and Ms I.van Dijk) for their excellent work in preparing and organising the Workshop. The Council of Europe would also like to express its thanks to the Rapporteur General (Professor Dr. L.F.W. de Klerk), the lecturers, and to the group chairmen and rapporteurs. The editing was done by Ms May Young, Ms Marie Thomas and Ms Pamela Munn, from the Scottish Council for Research in Education. They, too, deserve a warm word of thanks.

Michael VORBECK
Head of the Section for Educational
Research and Documentation

Strasbourg, 5 February 1987

PART 1:
REPORTS AND COMMISSIONED PAPERS

1.1 READING AND WRITING:
A (META) COGNITIVE VIEW

by

Prof. Dr. Len F.W. de KLERK, Rapporteur General

1.1.1 INTRODUCTION

Writing can be considered as the most important invention in the history of mankind. Without a written language, a highly civilized culture is unimaginable. Reading and writing are vital skills which are necessary for adequate functioning in every civilized country in the world.

It is generally accepted that the development of literacy in the native language is a very important objective of elementary education. However, there is a serious debate as to what must be the ultimate goals of the teaching of reading and writing. Some schools seem to stress the learning of basic skills (such as decoding and spelling). Other schools pay more attention to the functional and communicative aspects of reading and writing.

One of the main reasons for such a debate, which sometimes is quite feverish, is that the topic of reading and writing has attracted considerable interest from scientists of different disciplines. Besides policy makers, we find, among others, linguists, sociologists, and psychologists. Even within a particular discipline there may be several specializations. Among the supporters of the linguistic approach, for example, there are educational, socio-, psycho-, and anthropological linguists, each of them having his own ideas about how and what to teach with regard to reading and writing.

In this general report of the workshop on reading and writing, the (meta) cognitive view has been stressed.

1.1.2 TECHNICAL READING:

1.1.2.1 Decoding and recoding

School reading is a broad curriculum area, involving many instructional objectives. With respect to these objectives, a distinction can be made between —

what is referred to as — technical reading and reading comprehension. The main purpose of the teaching of technical reading is to promote learning to translate or decode written symbols into the oral language the child already speaks. This decoding process stands at the core of reading.

In order to be able to learn to decode printed words, the child's oral language skills must be relatively well developed.

Each language has a restricted number of phonemes. By using rules for combining these phonemes it is possible to produce a great many words. Similarly, rules for combining words make it possible to produce a virtually unlimited number of sentences.

Within the context of speaking and listening the child must have sufficient knowledge about these rules. He must know what sounds or combination of sounds belong to the oral language (phonology); he must know how words are put together in sentences (syntax), and what words stand for (semantics). The learning of the various letter-sound correspondences is dependent on certain pre-reading skills. Not only must the spoken word be available to the beginning reader, he must also be able to identify letters and letter configurations. Although there is a tendency to see an object's shape as unchanging regardless of the visual orientation we see it in, the child must learn, however, that a "b" is not a "d", nor a "p" a "q". When a set of objects is rearranged, it still remains the same set of objects. Yet, the combination "au" is not the same as "ua". Pre-reading skills such as discrimination and identification of letters, left-right processing of information, and the associations of symbols with sounds, must be sufficiently mastered before formal teaching of technical reading.

1.1.2.2 Comprehension

Technical reading focuses on the identification of the meanings of individual words. It is assumed that the meanings of words are stored in that part of the long term memory that is called the mental lexicon. The entries in this lexicon are not arranged alphabetically. There are essentially two mechanisms by which an entry can be accessed. One is to use the visual pattern of a word as a code, and the other mechanism is to sound out the words. Here it is through vocalization that the meaning becomes activated.

In skilled reading, the visual code is faster and will be first to reach the lexicon. However, when a word is not immediately recognized or difficult, then the sound code might be more adequate. In both cases, lexical access will lead to the activation of literal comprehension of words.

Literal comprehension is one of the objectives of the teaching of reading. This is sufficient for reading tasks such as looking up departure times in a railroad time schedule or telephone numbers in a telephone book. The ultimate goal of reading is, however, comprehension in the sense of understanding utterances or drawing inferences from them. Thus, comprehension may even involve going beyond the information literally or explicitly stated.

1.1.2.3 Reading as an interactive process

According to Weiss (1986), learning to read is a developmental process, whereby a number of stages can be distinguished. In the first stage, the child discovers that there must be some relationship between the oral language and what is presented in the form of prints. Gradually he learns to recognize the shapes of letters and the sounds they represent. In the second stage, the child starts reading by identifying the meaning of each word in a text, one at a time. Later on, as his vocabulary increases, the reader often uses only part of the visual information. He guesses the meaning of the words on the basis of his knowledge of the writing system as well as of the spoken language. The extent to which phonetical codes may guide his reading in this stage depends on the particular language.

In the third stage the surrounding context of the sentences comes to play an important role. The reading process probably begins as a bottom-up process, starting with the identification of letters and ending with the understanding of the meaning of whole passages. According to Weiss (1986) the reader will be guided more and more by his expectations about the text. In fact, reading becomes an interactive process (rather than a pure top-down process). In understanding a sentence, for instance, sometimes after getting just a few words, the reader may jump to what he thinks the entire sentence means. The few words may activate a whole scheme that is stored in his memory and that is used to help sentence understanding. Such schema guidance is very common, which means that language and memory are closely related in understanding both written and oral language (see also Schank, 1982).

1.1.2.4 Reading and problem solving

Failure to achieve the standards of performance for each of the various stages of the reading process is generally recognized as being a major problem in most primary schools in Europe. The information processing model of reading — as previously outlined — suggests that problems can occur in one or more component reading skills. There is evidence suggesting that skilled and poor readers may differ on each of these skills. This may be inferred from the experiments that have been discussed by Dehn (1986). This author has conceived of reading as a matter of problem solving. According to this view, trying to identify the meaning of a single word is a problem, as is trying to understand what a writer is communicating.

Good and poor readers do not appear to differ so much in the number or type of errors they make, but rather in the way they correct them. The good readers seem to be more flexible and accurate in correcting their errors.

From Dehn's (1986) review, it can be concluded that proficiency in decoding and recoding is needed in order to carry out higher order "meta operations". She did not find, however, that poor readers' decoding and recoding skills are not as accurate as skilled readers. Rather it appeared that both bypes of readers differ in the way they use strategies to attack their problems. Good readers are better able to develop and use heuristics — selective searches that look at those aspects of the problem that are most likely to produce a solution. This is in line with the classical work of Duncker, who found that problems often are reformulated into smaller subgoals. From this work it also appeared that hints may be very helpful in finding the correct solution. Dehn (1986) found that good readers make better use of such hints than poor readers do.

For the more advanced level of formal reading instruction, it should be taken into account that reading may be done for many reasons. To ensure that the goals of reading are met optimally, the meta-operations are of great importance. These refer to the reader's cognitive processes such as goal setting, strategy selection, evaluation, and remediation (see also Gagne, 1985).

A skilled reader sets a goal and selects a reading strategy. When the goal is, for example, getting an overview of a newspaper article, then skimming the text might be an adequate strategy. When directions for use must be understood in order properly to perform a series of operations, then the task requires literal comprehension for which reading word by word might be an adequate strategy.

A skilled reader will check from time to time whether his goals are being met. If not, then he must use some strategy to remediate the problem, such as re-reading or asking for help. This means that the reader must be able to recognize his problems and that he must know how to deal with them. The research, discussed by Dehn (1986), suggests that the heuristic approach that has been recommended, may help the reader to plan and to monitor his cognitive activities (such as planning, strategy selection, evaluation, and remediation) while he is performing the reading task. This is important, because skilled readers appear to be better at monitoring and controlling their meta-operations than less skilled readers (see also Palincsar & Brown, 1984).

1.1.3 COGNITIVE LEARNER CHARACTERISTICS

1.1.3.1 General

Generally speaking, it can be said that a teaching learning situation has to be designed in such a way that it will maximize the performance of the learner along prescribed dimensions and on specific criteria. The question is, however, to what extent and in what way a teacher must take account of the fact that individuals differ from each other in their abilities to read. Usually, difficulties occur when the child has been undergoing formal reading instruction for some time. It would be better, of course, if such difficulties can be predicted. The question is what the relevant initial conditions for reading instruction are. Which pre-reading skills, prior experiences, interests, beliefs, expectations and attitudes are the learners supposed to bring to the reading task? On which basis should instruction be built?

One way to get an answer to this question is to conduct fundamental research. An instance of this approach has been provided by Mommers (1986). The central question underlying his research project is which skills the reader must bring to each stage of the reading process. The emphasis in this study was on the relationship between spelling, decoding, and reading comprehension.

Mommers (1986) has administered a number of tests. Analyses of the test results showed that three different factors could be distinguished: a general ability factor and two more specific factors (i.e. a non verbal IQ factor and an auditory factor).

The general ability factor appeared to be a rather good predictor of reading comprehension achievement after some months of formal reading instruction. Later on, however, the influence of the general factor on reading comprehension decreased.

The auditory factor was positively correlated with both decoding and spelling. As formal instruction proceeded the influence of speed of decoding on reading comprehension increased, reaching the level of significance at the period of transition from beginning to skilled reading.

In general, the results are in accordance with the view that reading is a complex developmental process, involving different component skills. During formal instruction, the relationships between these component skills may change. This means that the impact of each component on skilled reading performance may be different at different points in time.

The importance of Mommers' (1986) approach is that it may provide useful information for the construction of diagnostic instruments. With the aid of these instruments it is possible to predict reading difficulties at various stages of the reading process. Prediction of reading difficulties is a *conditio sine qua non* for prevention. When a teacher knows which prerequisite skills are underdeveloped or are showing deficiencies, then training programs can be used to remediate the particular problems. The key to the success of reading instruction is that the readers can be motivated and helped to correct their difficulties at the appropriate points in the learning process. This requires early diagnosis and the availability of adequate training programs. Fundamental research may — in the end — provide the materials to make such a key.

1.1.3.2 Learner characteristics and reading materials

The fact that the reading process occurs according to certain stages has also been emphasized by Jansen (1986). Three stages are distinguished by this author, which run parallel with what we have referred to as decoding, literal comprehension and inferential comprehension, respectively. In fact, these labels can be seen as general indications of instructional goals. In planning for teaching, they provide a guide for choosing subject matter content and for sequencing topics. According to Jansen (1986), the content is only one aspect of the reading material.

Two other aspects should be taken into account as well: the language that is used (e.g. concepts and words) and the visual appearance of the materials (e.g. pictures, colours, letter types, etc.). These three aspects, together with the three developmental stages that have been distinguished, constitute a 3x3 matrix. This matrix can guide the selection of materials to be employed in the actual teaching process. It is a device to adapt instruction to the changing state of the learner: to his level of reading performance, his cognitive abilities, and his interests and motivation, at any particular point in time.

1.1.3.3 The reader's (pre) conceptions

In general, it can be said that a teaching/learning situation is designed to advance the learner from one level of achievement or competency to another. This means that the teacher always assumes that the learners possess some entry skills or readiness. These skills are usually defined in terms of developmental (cognitive) abilities and/or motivation. However, Olsson and Dahlgren (1986) have focused attention on another personal variable: the child's conceptions of the learning task. By using qualitative methods of investigation, these investigators were able to collect information about the children's (pre) conceptions of reading. On the basis of their conceptions it was possible to divide a number of pre-school children into two groups. The criterion was the "awareness of written language". Children who were aware of both the function of reading and the reading process itself were referred to as "being aware of written language". Children who were unaware of either one of the two aspects or both were labelled as "being unaware of written language".

These conceptions were "measured" just before formal instruction. After one year of reading instruction, information about the children's reading performance was collected. The results of this study showed that there is a rather high correlation between children's (pre)conceptions of the reading task and their reading performance. Obviously, the children's (pre)conceptions form another factor besides their basic cognitive skills which affect their degree of success at reading. Unfortunately, the sample was too small to know whether the predictions are sufficiently accurate to be useful. But this line of investigation is very promising, the more so because the authors have also indicated how the (pre)conceptions can be improved. They propose to use what they refer to as "metacognitive conversations". The teacher (or adult) should actively participate in such discussions, trying to make the children aware of why they should learn to read, and how reading will lead to success. These are assumed to be the key elements in obtaining good results.

1.1.3.4 Differentiation and aptitude treatment interaction

In the above, attention was focused on the (initial) conditions of reading instruction. It can be concluded that there are many factors that may influence reading performance. These factors include, besides the child's cognitive abilities (or disabilities), his pre-conceptions, prior experiences, attitudes, needs, etc.

These are personal variables, which are the result of a continuous interaction with the environment. Although it is often difficult to assess these variables, instruction would be largely wasted if it should not take them into account.

Usually, all children receive the same instruction. However, it is also possible that a child needs a special instructional treatment because he has some cognitive deficiences. In that case the teacher should take account of the fact that children with different abilities must be treated differently.

This approach has been advocated by Janssens and De Corte (1986). In Belgium, a package has been developed for (Flemish) language instruction in the primary school. For each topic, three different texts were constructed; a basic text for normal classroom instruction, a simpler version for the poor readers and an enriched programme for the proficient readers. This variety of text versions made it possible to accommodate to individual differences, but also to increase the motivation of the readers.

Every new topic began by focusing on common relevant experiences. By discussing these experiences, the teacher attempts to provide the learner with what Mayer (1975) termed a broad receptive set. After the oral discussion, written information is presented that is related to the particular topic. The receptive set and the continuous interaction between oral and written language are supposed to promote both reading competence and motivation.

After a few lessons, tests were administered to determine the children's progress. The poor readers were assigned to the simpler version of the text, whereas the skilled readers got the enriched version. Because the children were helped when and where they had difficulties in reading, this method of differentiation proved to be very successful. The children seemed to achieve better reading results, and they showed a high motivation for reading. Although this was also true for the relatively poor readers, the results indicated that, as time went by, the differences between the poor and skilled readers increased. This means that the skilled readers attained very high standards of reading performance.

For many teachers it appeared to be hard to manage the method of differentiation. Via in-service training and by providing the teachers with a good manual, containing many useful tips, the developers of the differentiation programme have tried to solve this problem.

According to this approach — which can be referred to as compensatory teaching — all children start with the same instructional method irrespective of possible cognitive deficiencies. For some children the particular method may be adequate, but for other children it may not be. With one general method for the whole class, these latter children may show difficulties. Often they even become labelled as retarded children. To remediate their difficulties it will be best to use a method of instruction that is adequately suited to their cognitive abilities. The same holds true for the more proficient readers. For them a general method might cause boredom, carelessness, or de-motivation, resulting in a reduced achievement.

11

The method of compensatory teaching can be further improved when it is based on research that is designed and carried out according to the "aptitude treatment interaction" (ATI) paradigm. The rationale for this type of research is that no one instructional method can be optimal for all pupils. Instead, one instructional method may be optimal for one group of pupils and an alternate method may be ideal for another group. ATI-research may reveal the relevant cognitive skills and abilities that must be taken into account when the method of compensatory teaching is used. Moreover, the effectiveness of each mode of instruction can be investigated in relation to the cognitive characteristics of the pupils. In planning ATI-research it might be profitable to follow Tobias (1982) advice not merely to "vary different instructional methods and monitor achievement or other outcomes. In addition, the types of cognitive processing activities that occur while students are working on instructional material should be studied". This appraoch will even be more powerful when it also includes metacognitive skills such as predicting, planning, checking and monitoring. In our opinion the time is ripe for combining ATI and the metacognitive psychology of information processing. Especially the development of training programmes that are needed for compensatory teaching will benefit from this combination.

1.1.4 WRITING

1.1.4.1 General

School writing has received relatively little attention. In some ways it is the opposite of reading. Reading is the understanding of ideas which are presented in printed form, whereas writing is the production of ideas in the form of prints.

Both reading and writing can be conceived of as complex developmental processes. The most basic component of writing is the motor skill of holding a pencil and producing letters. In the first, cognitive stage of the motor learning process, writing requires a great deal of attention and practice. Here, the process can be stimulated by teaching the learner how to analyze the structure of each letter (Pantina, 1957) and by providing feedback. Several studies suggest that skilled writing is largely guided by an abstract representation, or motor programme, that is stored in the writer's long term memory. For each letter, the characteristic pattern of up- and down strokes seem to be the relevant information that is contained in the motor programme. Other parameters, such as accuracy and speed, will be specified during the execution of the writing movement (Thomassen, Van Galen & de Klerk, 1985).

In a latter stage of the writing process, the writer pays more attention to spelling, grammar and punctuation while he is writing. In the end, writing becomes a complex process of generating text. This process starts with a mental representation of the topic and the goal, and ends with the production of a written composition. The essence of this ability is to generate and to translate ideas. This, however, is not a straightforward process. It is more likely to assume that it is an iterative process, starting with the generalization of ideas, followed by a writing phase which may also include evaluation and revision. The various component skills, such as generating, planning, and organizing ideas, translating these ideas, and reviewing and revising the text as it is produced by the writer, may occur either before, during or after writing, in any order. In fact, it requires a person to have a goal and to search for means to reach that goal. Once the goal has been set, there are many strategies that writers can use to reach it. However, the problem that is to be solved by the writer can be conceived of as an ill-defined problem. It is always very difficult to formulate good criteria that can be used to judge whether a solution is correct. It is also difficult to give adequate directions as to how to solve the problem. Strategies for solving such problems should include breaking the problem into several sub-problems and (re)organizing the whole situation.

Studies comparing good and poor writers reveal that they differ in the quality of the representation of the writing problem, and also in the organization of their knowledge with respect to the problem. These two aspects may affect the solution of the problem. We agree with Gagne (1985), who states that much research is needed to understand the problems of the poor writers. "We do not know, for example, the mechanisms by which problem representation and knowledge organization operate during problem solving. The coming years should give us a better understanding of problem solving and how it can be improved during the school years".

A problem solving approach to writing avails itself to large scale treatment within a European context, because its methods need not be bounded by the frontiers of specific languages. Far from being diverse, such a large scale approach may be disciplined by means of agreements between the research groups that are now already in existence, but not connected with each other through joint programmes. The Council of Europe seems to be the most likely candidate to initiate and coordinate such a large scale European research project.

1.1.5 CONCLUSIONS

From the above, a number of conclusions can be drawn which are worth noting. First of all, it can be said that a number of investigators have stated that reading is to be conceived of as a developmental process. Often, three different stages are distinguished. In the first stage, emphasis is on decoding. The second stage is a transient stage to the third, where the focus is on either literal understanding or inferential comprehension. However, there is no agreement whether formal reading instruction must follow these stages (bottom up) or that one has to start with the functional and communicative aspects of reading (top down).

Secondly, it appears that there exist large individual differences with regard to the sort of knowledge and experiences the child brings to the teaching/learning situation. The relevant personal variables are not limited to cognitive abilities (or disabilities) only. Variables such as attitudes, beliefs, motivation, experiences and preconceptions should also be taken into account. These types of variables are no stable personality traits, but rather a result of continuous interaction with the direct environment of the child. To encourage positive attitudes, beliefs and motivation, and to build instruction on existing experiences and (pre)conceptions should be major objectives for teachers. Only if reading and writing have relevance and purpose for the learner, may teaching then be optional.

Third, the child must learn to read for a variety of purposes. This requires the learner to set goals, to plan, and to see if the goals are being reached. These skills are studies within the context of metacognition and refer to the child's awareness and control of his own cognitive processes. Reading instructors should be more concerned with teaching and training of these metacognitive skills in addition to direct content instruction, which is usually their main concern. Children must be encouraged to become active participants in the process of learning to read. This is especially true when objectives such as comprehension and thinking (in the sense of drawing inferences) are included. In such cases, the learner should be encouraged frequently to ask himself questions such as: "Do I understand it?"; "Can I give a good example of it?"; "What else do I need to know?"; "Is this strategy working for me?". Emphasis on such metacognitive activities will lead to an improvement of reading performance.

Fourth, reading appears to be an interactive process. In fact, as a word (or sentence) is read, a mental representation of its meaning is formed in the short term memory. This is a temporary storage device which serves to mediate between long term memory and the execution of reading processes. Two key assumptions seem to be valid: (1) that reading is an interactive process, largely guided by the expectancies the reader has about the meaning of a word or a sentence (or even a whole passage) on the basis of what is stored in his long term memory, and (2) that this memory can be reached either directly, through a visual path, or indirectly through a sound path. Direct access will be promoted best by the whole word method, whereas indirect access is supposed to be emphasized by a more phonetical method. Nowadays, the so-called dual access hypothesis is becoming increasingly popular. According to this hypothesis, comprehension can be either direct or indirect, depending on the proficiency level of the reader and on the nature and difficulty of the reading materials. A combined approach, which is also in accordance with the interactive view of the reading process, seems to offer the best prospects for the teaching of reading comprehension.

A fifth conclusion is that it is not quite clear to what extent functional illiteracy, as it has been called, is recognized as being a social problem in every country of Europe. Though the concept is not clearly defined, it appears that the percentage of functional illiterates differs somewhat from country to country within Europe. But stating the problem does not guarantee a solution, nor will direct attack be the best way of arriving at it. What is needed is a solid and reliable knowledge base. This knowledge can be obtained from all available sources, e.g. fundamental research, theoretical formulations, and certainly also everyday observations. This information must be put together to form the ground for programmes, aiming at either prevention or remediation of reading difficulties. But equally important would be the need for a continuous process of testing these programmes in response to new evidence and, of course, feedback from the field. As a result of such an interaction between theory and practice, programmes will be so well developed in the near future that literacy problems will occur far less frequently than they appear to occur today.

Research on reading and writing has made significant but limited progress in the solution and prevention of difficulties. Analysis of the current status of this research in Europe suggests a need for co- ordination of our research efforts. We need to abandon a piecemeal approach. This can be stimulated, for example, by organizing European workshops on reading and writing, say bi-annually. Such workshops offer the possibility for researchers and policy makers to meet each other and to discuss the problem of (functional) illiteracy, which probably is a serious problem in most countries of Europe.

An important aspect of the whole issue is how to deal effectively with large individual differences in reading and writing capabilities. This is not a new problem, but a problem that has taken on new urgency. In order to be effective, research requires coordinated efforts on a vigorous European scale. Although the problems entailed in conducting such international large scale research programs cannot be over-estimated, both the practical and theoretical benefits that would accrue warrant the expenditure of these efforts. Let us hope that this will be challenging enough for the Council of Europe to take further initiatives.

REFERENCES

Dehn, M (1986) Investigations in learning to read and to write in class 1 as foundations of teaching methods for elementary schooling. Paper presented at the Educational Research Workshop on Reading and Writing Skills in Primary Education: Tilburg.

Gagne, E D (1985) *The cognitive psychology of school learning.* Boston: Little, Brown and Company.

Jansen, M (1986) Reading — a developmental skill? Paper presented at the Educational Research Workshop on Reading and Writing Skills in Primary Education: Tilburg.

Janssens, A and De Corte, E, (1986) Differentiated early reading instruction: experiences with a Flemish teaching-learning package. Paper presented at the Educational Workshop on Reading and Writing Skills in Primary Education: Tilburg.

Mayer, A E (1975) Information processing variables in learning to solve problems. *Review of Educational Research, 45,* 4, 525-541.

Mommers, J C (1986) The relation between decoding skills, reading comprehension and spelling in the first three years of primary school. Paper presented at the Educational Research Workshop on Reading and Writing Skills in Primary Education: Tilburg.

Olsson, L E & Dahlgren, G (1986) The child's conception of reading. Paper submitted to the Educational Research Workshop on Reading and Writing Skills in Primary Education: Tilburg.

Palincsar, A S & Brown, A.L. (1984) Reciprocal teaching of comprehension-fostering and comprehension-monitoring activities. *Cognition and Instruction , 1,* 117-175

Pantina, N S (1957) De wijze van oriëntering in de taak bij het aanvangsschrijfonderwijs. In C F van Parreren & J A M Carpay (eds) (1972) *Sovjetpsychologen aan het woord.* Groningen: Wolters Noordhoff.

Schank, A C (1982) *Dynamic Memory,* New York: Cambridge University Press.

Thomassen, A J W M,Galen, G.P. van and Klerk, L.F.W. de (eds) (1985) *Studies over de schrijfmotoriek: theorie en toepassing in het onderwijs.* Lisse: Swets & Zeitlinger.

Tobias, S (1982) When do instructional methods make a difference? *Educational Researcher, 11,* 4-9

Weiss, J (1986) The three stages of learning to read. Paper submitted to the Educational Research Workshop on Reading and Writing Skills in Primary Education: Tilburg

1.2 A SUMMARY OF RECENT RESEARCH ON READING AND WRITING IN PRIMARY SCHOOLS IN THE UNITED KINGDOM

by

Anne SANDERSON, United Kingdom

1.2.1 SUMMARY

Many current research projects on language teaching and learning in primary schools acknowledge the inter-relationship of the different language modes. Links between reading and writing clearly emerge, and reports suggest common factors in the development of and progress in literacy. Detailed analyses and observations of successful early and later avid readers highlight the importance of young children's early encounters with print in the home and later in the school environment. It is clear that, if reading and writing have relevance and purpose for the learners, and if conditions help them to perceive themselves as readers and writers, then positive attitudes towards literacy are likely to develop. National surveys suggest that negative attitudes towards reading and writing correlate with poor performance, and therefore the encouragement of positive attitudes is a major objective for teachers.

Teacher responses are found to vary considerably, and undoubtedly the nature of these affects learners' views and perceptions of the nature of reading and writing. When pupils are given a more realistic view of what is involved they develop greater expectations of success. Evidence suggests, however, that many pupils throughout the primary age range have a narrow view of literacy with exaggerated aims of neatness, correctness and surface features rather than interrogation of text. They have, in fact, been given limited criteria for making judgements about print.

The research summarised in this paper is by no means exhaustive. A growing body of teachers and researchers is observing current practice in the teaching of reading and writing in order to review the practical possibilities for developing a broader and richer curriculum.

1.2.2 INTRODUCTION

This paper is an attempt to summarize aspects of reading and writing in primary schools, currently of interest to educationists in the UK.

The range and variety of research into these areas has expanded considerably over the past five years, covering national and local surveys, experimental design, action-research and ethnographic studies of individual schools or classrooms. Many teachers are collaborating with researchers to examine how children respond to aspects of language teaching through observation of the processes by which pupils become literate. Information gained from such research is increasingly focussing attention both on the processes of reading and writing and on the vital contributions made by parents to this area of learning.

1.2.3 RESEARCH INTO READING AND WRITING IN PRIMARY SCHOOLS

1.2.3.1 General

The most extensive and highly publicised research into the language of pupils in primary and secondary schools is that carried out by the Assessment of Performance Unit (APU). In designing an assessment framework (for the monitoring of language performance), the following assumptions were made:

- that using language is a co-operative activity
- that activities should reflect real purposes
- that using language in school involves the association of different language modes
- that different varieties of language are used in different circumstances
- that communicative acts, spoken or written, have common features of components
- that communication in different modes involves different demands, and performance may be affected by these demands
- that performance in language is task-related
- that performance and attitude are inter-related.

1.2.3.2. Reading in the primary schools

The assessment of reading surveys was based upon three broad categories of material: works of reference, works of literature, and materials which pupils might use for practical purposes (Appendix 1). The focus and scope of the text were varied, and tasks determined which reading strategies pupils needed to adopt.

Assessment was based upon:
1   Pupils' responses, coded and assessed on a pro-forma record sheet.
2   A selected sample of responses analysed in depth to highlight processes and

strategies differentiating the most successful, and least successful, groups of readers.

The reading surveys provided evidence that although few 11 year olds in the sample had problems decoding print, many were unable to extract meaning through interrogation of text.

Common difficulties in low performance pupils were identified by the research team. These included inability to locate information from reference books and misinterpretation of text and stylistic variations. Pupils' own pre-conceptions or misconceptions of what they read frequently obscured the intended meaning.

Attitudes to reading were investigated and based upon:
- a view of the act of reading
- satisfaction gained from reading
- pupils' perceptions of themselves as readers
- reading in the home environment.

Low performance in reading and negative attitudes were found to correlate.

The following research projects are based upon observations of children involved in the reading process and of home factors which contribute to success in literacy.

Ingham (1982) investigating *home influences and early reading experiences of 11 year olds* reported that:

reading attainment was not associated significantly to social class, ethnic origin, intelligence or family size;

successful early readers and later avid readers had all experienced frequent story reading and interaction with varied text;

the successful readers had learned the process of reading through observation of and interaction with parents and teachers.

A study by Hynds (1984) of the *home experiences of early successful readers,* highlighted the following optimum conditions for reading in the home:

the main purpose of reading   – enjoyment (not to learn to read)

children's book experiences   – from the age of 2 or earlier, experiencing 200 or more books a year

                                            – rich book language with lively, imaginative or humorous content

children's reading behaviour   – a core of favourite books, constant observation of 'readers', children rarely required to
read to parent, parent reads to child.

It was observed that these conditions are often in sharp contrast to conditions in
many schools.

A single child's early interaction with print in a literate home environment was
monitored by Payton (1984) and continued through to her accomplishments in
relation to the demands of the reading process. The child was found to predict,
analyse and accommodate information and knowledge about literacy through
consistently initiated and sustained cognitive activity. Thus she developed the
complexities of both reading and writing.

Dombey's (1985) case study investigated a *three and a five year old interacting in
story sessions* at home and in nursery school, and emphasized the important
influence of the adult. Analysis in close linguistic detail of frequent recorded
story-reading sessions revealed that, through such exchanges, children are
capable of construing and constructing many of the forms of written language
when they are embedded in conversations focussing on narrative. Participation in
story reading, the study shows, is a major facilitator in the transition from speech
to writing and in acquisition of the functions of written language. This is
emphasized as being highly relevant to the reading process.

1.2.3.3 Aspects of reading in primary schools

Ingham (1985) investigating how *reading* may be made more interesting,
meaningful and accessible to minority groups in school, designed a project
involving research teams in three London boroughs. Pupils were interviewed
about their language use, reading habits and story-telling in the home. Through
home case studies, parental involvement in their children's reading was
facilitated. This involved parents in story-telling, use of minority groups' folk
tales in school and the production of dual language reading materials. Such
community involvement was found to increase children's confidence in reading
and self-esteem through legitimizing stories they had heard in the home.

In a study of *seven year olds' use of inference in understanding stories,* Oakhill
(1984) compared children of the same reading accuracy but with different
comprehension scores (Neale). The more advanced comprehenders were better at
answering questions from memory and inferring from the text. The least
advanced comprehenders performed equally well on literal questions when the
text was available, but failed to infer from the text, thus emphasising that reading
is more than decoding print, and that teachers need to encourage readers to relate
their knowledge and experience to what they read, as do successful comprehenders.

Francis (1984) investigated *children's knowledge of orthography in learning to
read* throughout their infant schooling by contrasting actual spellings in text with

pupils' miscues. The study revealed that whilst children are reading aloud they acquire considerable knowledge of spellings independent of the explicit use of phonics by pupils and teachers. The main issue arising from this study is whether the visual, analytic and organising abilities children bring to bear on print are sufficiently recognised by teachers as a real basis for phoneme-grapheme understanding.

The effects of auditory organisation on later reading success were examined by Bradley and Bryant (1985). This research involved a sample of 400 'non-literate' four and five year olds. Experimental groups were exposed to frequent intervention activities involving rhyming experiences and alliteration in oral word play. Multiple regression analyses revealed that early skill in auditory organisation accounted for significant variance in children's progress in reading. Intervention strategies significantly affected later performance in reading. The potential for parents, exacerbating the reading process through rhyming and word play in the pre-school years, is emphasized through this research.

In her study of *children's oral reading development between the ages of five and eight* Perera (ongoing) is focussing on prosodic features and readers' progression from word by word emphasis through to fluency of expression and intonation. A prosodic analysis of the development of six children learning to read will provide an interesting body of information in a relatively unresearched area.

Responses of adults in the reading process are being increasingly considered as a vital factor in interest and successful reading.

Hall (1984) examined *the extent to which teacher behaviour and books transmit values about literacy to children.* Through a detailed analysis of six commonly used reading schemes in the country, he reached the conclusion that true literacy events were scarce and characters did not look happy when engaged in them, or they were restricted to the school situation. It was suggested that the messages conveyed to children were that reading is a marginal activity, highly school-based, not functional and not pleasureable. Hall suggests that where teachers demonstrate their own interest in reading and write for the children, the status of literacy is higher in the perceptions of the pupils.

Waterland (1985) proved that although parents are more effective in helping children to read, schools can facilitate continuity of approach through their reading provision. Her case study includes *detailed observations of young readers in school* and focusses on the 'apprenticeship' approach to reading. Interaction and interpretation of text was seen as a priority. Observations, based upon a Reading Behaviour Development Chart (Appendix 3) proved that the commitment of parents and teachers to this approach positively affected children's attitudes and interests in reading, and their perceptions of themselves as readers.

Examining the effects of teacher responses when learning children read, Campbell (1985) recorded 150 daily oral reading sessions of two six year old boys and teacher intervention/response in the process. Recordings were obtained in naturalistic classroom settings.

Teacher responses were categorised as:

non-response   – to a syntactically or semantically appropriate substitution.

providing word   – immediate 'telling' of the unknown word.

word cueing   – highlighting certain characteristics of the word in context.

The least effective response was to tell the reader the word, whilst the most effective was the word cueing move. Thus cognitive involvement in the cue systems of reading proved to be most helpful in subsequent meetings of the word.

Meek, *et al* (ongoing) are pursuing their researches in terms of 'transitions' or moves and changes made by children relating to their view of the task of learning to read, and the nature of their encounters with different texts.

Through recorded discussions and/or reading events, evidence indicates that pupils, besides reading a text, also 'read' the teacher, in order to adopt the expected reading behaviour of the class. The social and collaborative nature of reading is evident, through transcript analysis, and also that changes in perception of readerly behaviour can lead to success for previously failing pupils. The series of case studies includes children from 4 - 12 who appear to be adept at changing their views of reading and modifying them to suit pedagogic practices in various school situations.

Steirer (1985), through an *ethnographic study of teachers' reading assessment and judgement-making processes,* aimed to demonstrate the suitability of qualitative research approaches in the study of reading practice in schools. Two teachers were involved in the study. Observations through note-taking and tape recordings were made over a six month period. The focus is on the following themes:

(a) Reading judgements and reading programmes.
(b) The implications of children's oral and silent reading.
(c) Teacher definitions of the competent reader.
(d) Teachers' judgements and external authority in reading assessment (sociolo-gical tensions).

Information from the study highlighted the social influences in classrooms, on the reading process, and the variety of strategies used by teachers for assessing readers through oral and silent reading.

Research into young children's becoming literate frequently links the areas of *reading and writing* (Dombey; Payton; Hall; Raban; Bradley). Mays' (1985) research into *story writing by 9 year olds* revealed that pupils who produced stories, exhibiting a range and variety of narrative techniques, were identified as having been fluent readers at the age of 7. Mays concluded that through reading and listening to stories, narrative structures are internalised, thus enabling pupils to create their own. Empirical data demonstrates that writing in response to a model provides strong evidence for close links between reception and production of stories. This supports the theory that frequent story reading is a vital activity in the development of writing.

1.2.4 WRITING IN THE PRIMARY SCHOOL

1.2.4.1 The APU (1979 - 1983) surveys of pupils' writing were based upon a threefold framework for assessment:

- writing tasks varied in nature and style (Appendix 4)
- attitudes to writing through open ended questions and statements
- work sampling from normal classroom experiences.

Thus 11 year olds were assessed on how they responded to various written tasks, what they produced in school and their attitudes to themselves as writers and their own work.

In the surveys children's writing was marked in two ways: impression marking (1-7 scale), a rapid but comprehensive judgement of the text, and analytic marking (1-5 scale), close and detailed reading of the script. The analytic marking scheme assessed writing with respect to content, organisation, appropriateness and style, knowledge of grammatical conventions and knowledge of orthographic conventions.

Some important findings emerging from this research are that

about 3% of 11 year old pupils are in great difficulty with writing;
about 60% produced interesting and legible work;
pupils' own judgements on how to improve writing is based upon surface features;
pupils' writing is influenced by the way they speak;
pupils' ability to write purposefully and well is dependent upon specific features of the task;
negative attitudes towards writing are associated with poor performance (more boys than girls show negative attitudes);
girls achieve higher marks for writing than boys.

The National Writing Project (1985-1988) (Czerniewska) is at present based in 20 local education authorities in England.

Many of the recommendations arising from the APU writing surveys are being researched and implemented into action-learning in schools. Research teams, co-ordinators, teachers and pupils are collaborating and investigating aspects of writing development and the process of writing between 5 and 16. Five major areas are emphasised in the designs:

how children learn to write and use writing for learning;
ways of providing *real* purposes for writing;
the beginnings of writing;
pupil perceptions of writing;
continuity between phases of education.

The Project strands vary in focus, but throughout this SCDS research project area teams are working alongside teachers and investigating specific aspects of the writing process.

Concentration on the beginnings of writing reflects growing concern that children, from their earliest encounters with writing, should see it as a meaningful activity. The *Making Beginning Writing Meaningful Project*, (Hall, 1986) aims to investigate:

the extent to which writing is a meaningful activity to children prior to, or reaching, full time education;
current provision in nursery and reception classes for making writing meaningful;
strategies and policies within these classes for making writing meaningful.

This two-year action-research plan involves researchers and teachers from a group of 12 schools working together, and begins the second phase of the National Writing Project. It will hopefully influence school policies for initiating young children into worthwhile writing experiences.

The Foundations of Writing Project, Jackson and Michael (1983-1986) was based in Scotland and originated from the examination of a large sample of writing from primary schools. This initial survey revealed that primary pupils had limited knowledge of the range and style of writing and had limited access to helpful criteria for judging their work.

Phase 1 – examined the possibility of teaching the higher order skills of writing, composition and expression separately from the lower order skills of handwriting, spelling, punctuation and grammar.

During the first three years of schooling, the following stages were experienced:

- pupils were encouraged to communicate through drawing and talk – teacher as secretary;
- narrative skills were incorporated into writing characteristics – letter formation began;
- children began to write;
- extension of narrative skills into themes, group work or teacher as scribe;

- personal writing to support a variety of experiences and extension of purposes.

Thematic writing, rooted in experience, resulted in the production of collections of writing in a range of styles and for varied readership.

A research project based at the University of East Anglia, Nicholls (1985) involved teachers and researchers in examining procedures to develop the writing of children between the ages of 5 and 9.

This *Beginning Writing Enquiry* group evolved a detailed observation schedule for observing young writers in five different schools (Appendix 5). Findings from this stage implied that the mechanics of writing were obscuring the composition and organisation of ideas. Further investigations revealed that children's key concepts about writing were mostly concerned with surface features. Close observations were made of how individual children approached written tasks in the classroom. Strategies to understand the basis of misunderstandings and errors evolved in the early stages of this research.

Through monitoring both the composing and performing aspects of writing, the project team produced a consistent account of the needs of young writers and the changing role of the teacher in this process. Although tentative models of writing were produced, further evidence is being collected by 15 teachers, all researching their own classrooms. (Appendices 6 and 7).

The effects of a real audience on children's writing was investigated by Thorogood (1985) over a 12-week period in school, and involved fifteen 8 year old pupils.

Group A spent extra time writing materials for younger children to read or have read to them.
Group B were given extra maths functions.
Group C received no extra input.

All the children wrote a story for younger children at the beginning and end of the project. Group A produced marginally more appropriate stories at the beginning of the project, but much higher quality at the end of the period. The most marked change in the experimental group, however, was in their writing behaviour and attitudes to writing. At the end of the 12 weeks they collaborated, planned, revised and wrote in their own time prior to publication.

There is growing interest in the development of writing in primary schools and many small-scale ethnographic studies are emerging.

Robson (ongoing) is investigating by case study *the development of writing in groups of children between the ages of 8 and 10*. The major focus is on detailed classroom observations and work sampling over the two-year period. An observation schedule designed to produce information on external influences on the writing process has been developed. The three children's spoken language, sociolinguistic features, attitudes towards literacy, and their perceptions of the writing task, are to be considered as possible influences.

Syntax, semantics, cohesion and discourse will be examined in comparison with the children's communicative competence. Through this study Robson aims to illuminate the complexities of becoming a writer and the problems some children experience as learners.

REFERENCES

Bradley L & Bryant P (1985) Reading skill in young children and the recognition of auditory similarities (SSRC) Department of Experimental Psychology University of Oxford.

Campbell R (1985) Oral reading and teacher instructions Unpublished PhD University of London Institute of Education.

Czerniewska P (Director) National Writing Project 45 Notting Hill Gate, London W22 5JB.

Dombey H (1985) Aural experience of the language of written narrative in some pre-school children and its relevance in learning to read — source missing.

Francis H (1984) Children's knowledge of orthography in learning to read British Journal of Educational Psychology, 54, 8-23.

Gorman T (1986) (APU) The Framework for the Assessment of Language.

Hall N (1984) Conveying the message that reading is necessary and pleasant, in D Dennis (ed) Reading; meeting children's needs. Heinemann Educational.

Hall N et al (1986) Making beginning writing meaningful NWP

Hynds (1984) Developments in reading and their implications for school and classroom practice. Avery Hall College. Thames Polytechnic.

Ingham J (1985) Reading materials for minority groups — Preliminary Study of Access to Ethnic Minority Materials. Department of Education, Middlesex Polytechnic.

Ingham J (1982) Books and Reading Development Heinemann.

Jackson W J & Michael W (1986) (Directors) Foundations of writing project, SCDC. Moray House College of Education, Holyrood Road, Edinburgh EH8 8AQ.

Mays S (1984) Literacy models and the structure of children's stories. Institute of Education. University of London.

Meek M et al (ongoing) Transitions — moves or changes made by children which are related to their view of the task of learning to read. University of London Institute of Education.

Nicholls J (1985) (Consultant) Learning to Write and Teaching Writing. 5-9 years. School of Education University of East Anglis, Warwick NR4 7TJ.

Oakhill J (1984) Inferential and memory skills in children's comprehension of stories. British Journal of Educational Psychology, 54, 31-39.

Payton S (1984) Developing awareness of print — a young child's first steps towards literacy. University of Birmingham.

Perera K (ongoing) The development of prosodic features in children's oral reading. Department of Linguistics University of Manchester.

Raban K B (1984) Observing children learning to read and write. School of Education University of Bristol.

Robson C (ongoing) A longitudinal study of writing development in a group of children between the ages of 8 and 10. Newcastle Polytechnic.

Steirer B (1985) Teachers evaluating children's reading — an ethnographic study of everyday classroom assessment practices for reading. Unpublished PhD Thesis University of Bristol.

Thorogood L (1982) Developing children's writing skills: a writing for others project School of Education, University of Reading.

Waterland L (1985) An apprenticeship approach to reading Thimble Press

APPENDICES

Appendix 1

A functional range of reading activities in which pupils might be engaged (APU)

1 Reading to give an overall impression of a single passage or chapter.
2 Reading of different passages to select information relevant to a particular topic.
3 Reading to expand upon information previously supplied.
4 Reading to follow a sequence of instructions.
5 Reading to identify the answers to questions by direct reference to a given text.
6 Reading to detect information implied in a passage.
7 Reading to interpret and evaluate a writer's assumptions and intentions and to show an awareness of the characteristics of different kinds of writing.
8 Reading for pleasure.

Appendix 2

Parental involvement in children's reading

Recent and relevant information and reviews of research into the growing area of parental participation in school reading programmes and implications for further developments.

Bloom W (1986) Partnership with Parents in Reading Hodder & Stoughton.
    An assessment of many of the current projects which actively involve parents in their children's learning to read.
Branston P and Provis M (1986) Children and Parents Enjoying Reading Hodder & Stoughton.
    A wide ranging summary of up-to-date research findings on parental involvement in their children's reading followed by practical advice on setting up such schemes.
Grigg S (1984) Parental Involvement with Reading.
    An experimental comparison of paired reading and listening to children read.
    Unpublished MSc (Ed.Psych) University of Newcastle upon Tyne.
Hannon P (forthcoming) The Belfield Reading Project - An evaluation of the Project five years after its instigation. Sheffield University.
Jackson A and Hannon P W (1980) The Belfield Reading Project Rochdale Belfield Council.
Topping K and Wolfendale S (1985) Parental Involvement in Children's Reading. Croom Helm.
    A review of past and current good practice in this field. Details of a wide range of schemes developed in local areas, given in a series of short contributed papers.

Appendix 3

READING BEHAVIOUR DEVELOPMENT   WATERLAND (1985)

- Child listens to story - watches pictures
   Child listens to story - observes text.
   Child listens to story - vocalises with adult.
   Child offers to read some or all of text at any level below.
- Child 'makes up' story - no pointing, word match or even accurate recall. May not even seem to look at print.
   Child 'tells' story - accurate, retelling, no pointing or word match. Aware of the text.
   Child tells story with finger. Accurate retelling of story, runs finger along lines, often matches beginning and ending of line with voice and finger.
   Child tells story with word/sound match. Tells accurately, complete word-by-word voice match. Very fluent.
   Child reads known words. A mixture of reading and telling. Child may comment 'I've got that word in my sentence maker.' Unknown words may be made up entirely, using no clue other than general

27

sense. Child's finger often stabs at known words, slides over unknown. Still very fluent.
Child reads known words; uses context, phonic clues, general configuration to decode unknown ones. Decoding often inaccurate, e.g. 'house' for 'home'. Reading may slow down to word-by-word rate.
Child reads known words, decodes unknown ones with accuracy, using context, phonics, and so on. Unknown words are far fewer and tackled with confidence. Reading may be basically fluent, slowing down at 'tricky bits'.

Appendix 4

Written tasks: Assessment of Performance Surveys of 11 year olds.

General purpose

To describe/observe
    Description of a memorable person or animal.
    Comparative description of insects/moths.
    Description of a picture.
    Description of a game.

To record/report
    Eye-witness account of a series of events.
    An account of something learned.

To plan/speculate
    Planning an experiment.
    Giving an account of an activity to be undertaken.
    Planning a room design.

To narrate
    A short story.
    Autobiographical anecdote about a past experience.
    Story completion (given a situation described and/or a picture).

To instruct/direct
    A poster announcing an event.
    An account of a skill.

To argue/persuade
    Persuasive argument based on personal opinion.
    Dramatised argument from a particular viewpoint.

To evaluate/review evidence
    Review of book/TV programme

To correspond/request
    Personal letter.
    Letter of Request.

To explain/reflect
    Explanation and justification of a rule or regulations.
    Explanations of reasons for selecting an experimental design.
    Explanation of reasons for liking a game.

To respond imaginatively/to express feelings
    Interpretation and response of poems.

Appendix 5

OBSERVATION SCHEDULE FOR BEGINNING WRITING ENQUIRY

a) Child No _____   b) Observation No _____   c) Date _____   d) Observer's Name _____

| Topic Source: | 1. Child chosen ☐ | 2. Teacher specified ☐ | 3. Talk mediated ☐ | 4. Multi-sensory ☐ |

FOCUS OF WRITING TASK

| Mode Intended | 5. Unspecified ☐ | 6. Story-recalled or original ☐ | 7. Report or description ☐ |

| Audience: 8. Unspecified ☐ | 9. Child chosen ☐ | 10. Teacher specified ☐ |

ILLUSTRATION: 11. Before writing   12. During writing   13. After writing

TEXT

| Pre-independent only: | 14. Teacher initiated ☐ | 15. Teacher prompted ☐ | 16. Teacher transcribed ☐ |

| | 17. Child initiated ☐ | 18. Child elaborated ☐ | 19. Child monitored ☐ | 20. Child copied ☐ |

CONSTRUCTION

All Writing:

| Planning talk: | 21. Before writing ☐ | 22. During writing ☐ |

| Doing writing: | 23. Letters ☐ | 24. Strings ☐ | 25. Words ☐ | 26. Groups ☐ |

| Vocalising while writing: | (27. Letter sounds ☐ | 28. Letter names ☐ | 29. Words ☐ | 30. Groups ☐ |
| | (31. Sound effects ☐ | 32. Commentary ☐ |

USE OF SOURCES

Code
v = vocabulary
i = ideas
r = reassurance
s = spelling
e = expression

Classmate	33 v	34 i	35 r	36 s	37 e

Teacher	38 v	39 i	40 r	41 s	42 e

Written text	43 v	44 i	45 r	46 s	47 e

Environmental	48 v	49 i	50 r	51 s	52 e

29

OBSERVATION SCHEDULE FOR BEGINNING WRITING ENQUIRY

OTHER ACTIVITIES:

Before Writing: 53. Management ☐ 54. Silent sitting ☐ 55. Diversion ☐

During Writing: 56. Management ☐ 57. Silent sitting ☐ 58. Diversion ☐

MONITORING OF TEXT:

Reading Over: (Aloud: 59. Part text ☐ 60. Whole text ☐

(Silently: 61. Part text ☐ 62. Whole text ☐

Discussion of text 63. With classmate ☐ 64. With teacher ☐

Alteration 65. With Spelling ☐ 66. Wording ☐ 67. Punctuation ☐

USE OF TIME

68. Before writing anything ☐ 69. Before starting text ☐

70. To finish writing ☐ 71. To finish the task ☐

Categories used in the schedule

Please note: Observations are recorded in the form of a 'running commentary' on what is happening every second minute.

Appendix 6

TWO TENTATIVE MODELS OF WRITING DEVELOPMENT

MODEL 1: TOWARDS ASSOCIATIVE WRITING

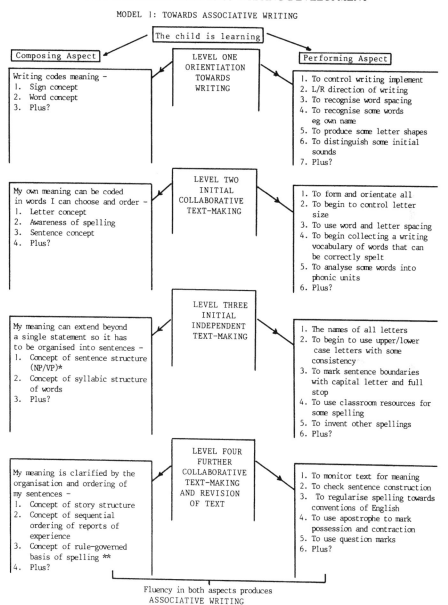

The child is learning

| Composing Aspect | | Performing Aspect |

LEVEL ONE ORIENTATION TOWARDS WRITING

Writing codes meaning –
1. Sign concept
2. Word concept
3. Plus?

1. To control writing implement
2. L/R direction of writing
3. To recognise word spacing
4. To recognise some words eg own name
5. To produce some letter shapes
6. To distinguish some initial sounds
7. Plus?

LEVEL TWO INITIAL COLLABORATIVE TEXT-MAKING

My own meaning can be coded in words I can choose and order –
1. Letter concept
2. Awareness of spelling
3. Sentence concept
4. Plus?

1. To form and orientate all
2. To begin to control letter size
3. To use word and letter spacing
4. To begin collecting a writing vocabulary of words that can be correctly spelt
5. To analyse some words into phonic units
6. Plus?

LEVEL THREE INITIAL INDEPENDENT TEXT-MAKING

My meaning can extend beyond a single statement so it has to be organised into sentences –
1. Concept of sentence structure (NP/VP)*
2. Concept of syllabic structure of words
3. Plus?

1. The names of all letters
2. To begin to use upper/lower case letters with some consistency
3. To mark sentence boundaries with capital letter and full stop
4. To use classroom resources for some spelling
5. To invent other spellings
6. Plus?

LEVEL FOUR FURTHER COLLABORATIVE TEXT-MAKING AND REVISION OF TEXT

My meaning is clarified by the organisation and ordering of my sentences –
1. Concept of story structure
2. Concept of sequential ordering of reports of experience
3. Concept of rule-governed basis of spelling **
4. Plus?

1. To monitor text for meaning
2. To check sentence construction
3. To regularise spelling towards conventions of English
4. To use apostrophe to mark possession and contraction
5. To use question marks
6. Plus?

Fluency in both aspects produces
ASSOCIATIVE WRITING

* NP = Naming part
 VP = Verbal part
** That is to say, the child is learning that successful spelling is based on careful VISUAL attention to 'letter strings' (not merely on consulting one's ear).

31

Appendix 7

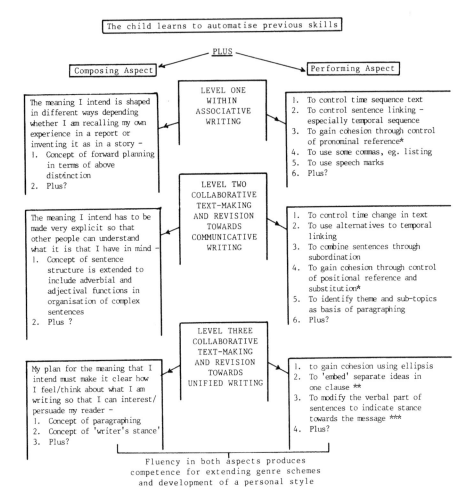

MODEL 2: DEVELOPMENTS FROM ASSOCIATIVE WRITING

The child learns to automatise previous skills

PLUS

Composing Aspect   Performing Aspect

LEVEL ONE
WITHIN
ASSOCIATIVE
WRITING

The meaning I intend is shaped
in different ways depending
whether I am recalling my own
experience in a report or
inventing it as in a story –
1. Concept of forward planning
   in terms of above
   distinction
2. Plus?

1. To control time sequence text
2. To control sentence linking –
   especially temporal sequence
3. To gain cohesion through control
   of pronominal reference*
4. To use some commas, eg. listing
5. To use speech marks
6. Plus?

LEVEL TWO
COLLABORATIVE
TEXT-MAKING
AND REVISION
TOWARDS
COMMUNICATIVE
WRITING

The meaning I intend has to be
made very explicit so that
other people can understand
what it is that I have in mind –
1. Concept of sentence
   structure is extended to
   include adverbial and
   adjectival functions in
   organisation of complex
   sentences
2. Plus ?

1. To control time change in text
2. To use alternatives to temporal
   linking
3. To combine sentences through
   subordination
4. To gain cohesion through control
   of positional reference and
   substitution*
5. To identify theme and sub-topics
   as basis of paragraphing
6. Plus?

LEVEL THREE
COLLABORATIVE
TEXT-MAKING
AND REVISION
TOWARDS
UNIFIED WRITING

My plan for the meaning that I
intend must make it clear how
I feel/think about what I am
writing so that I can interest/
persuade my reader –
1. Concept of paragraphing
2. Concept of 'writer's stance'
3. Plus?

1. to gain cohesion using ellipsis
2. To 'embed' separate ideas in
   one clause **
3. To modify the verbal part of
   sentences to indicate stance
   towards the message ***
4. Plus?

Fluency in both aspects produces
competence for extending genre schemes
and development of a personal style

*   The concept of cohesion is critical to writing development, this term and the
particular examples of 'pronominal' and 'positional' reference, substitution
and 'ellipsis' are all discussed in: Halliday, M. & Hasan, R, 1976
Cohesion in English, Longman

**   We all use this linguistic technique though there is no familiar term for it –
eg 'Yesterday I wrote to Hans my German pen-friend' = Yesterday I wrote
to Hans. Hans is my pen-friend.

***   This is really a reference to the linguistic concept of 'modality'.

32

1.3 READING DEVELOPMENT

by

Jacques FIJALKOW, France

1.3.1 SUMMARY

For some years the subject of "reading development" has again been of major concern. Discussions revolve around questions of examination failure and teaching methods revitalised by insights from the language sciences.

The way in which teachers effectively teach children to read is not yet properly known. There are no longer any "methods" in the conventional meaning of the term. The dividing line is really between classes using a textbook and the rest. Writing skills are neglected at the beginning of the primary school.

The more innovatory classes use children's literature, social texts, library-documentation centres. They use reading kits and place more emphasis on writing skills.

Three recent research projects are concerned with the problem of the factors affecting the acquisition of the written language in schools. Social background, national origin, age and sex are examined. One important fact is the convergence of results and, in mathematics, leading to the formulation of recommendations for improving the situation not strictly speaking in the sphere of the written language.

In the classroom, it seems desirable to emphasise the content rather than the technique of reading. Children's literature and texts to be read in connection with a project have the same objectives. Teaching kits and games enable verbal material from them to be used again. Writing skills should be widely used, in line with the structuralist approach to the recommendations for reading activity and its development. These measures require changes in the organisation of classes in order to develop individual and small group work.

In the schools, the establishment of a team of teachers, responsible for the first teaching stage over three years, is proposed. The team would have a regular time allotted for planning purposes. The selection of teachers responsible for the first year of compulsory education is a special problem. The school teaching team, self-selected to carry out a teaching project, would benefit from a head with an interest in project leadership, from in-service training oriented towards the team rather than the individual, and from the support of parents with a commitment to the project.

Within the school and its environment, the participation of parents and neighbourhood associations in the management of the library-documentation centre would help. Similarly, in-service training seminars bringing together teachers, parents, librarians and social workers in the school neighbourhood concerned would help to constitute the local network without which the school will remain an institution cut off from its environment.

1.3.2 INTRODUCTION

The subject of "reading development", after occupying the headlines some twenty or thirty years ago (when there were battles over the "global method"), had been forgotten. It was resurrected a few years ago and once again occupies the front of the stage.

Current interest in "reading skills" is widespread and no subject relating to school life arouses so much passion and energy.

The abundant literature proving this interest is mainly the work of teacher trainers and teaching movements or associations. Publications describing experiences based on school life are becoming more common. Published reports of discussions or of scientific meetings are no longer rare. The French Ministry of Education contributes to this current awareness through its activities.

If we try to stand back from the fray and define what is being discussed it seems possible to distinguish two main approaches.

The first, and the older one, concerns examination failure. It starts from the observation made twenty years ago by sociologists, and taken up since then and exploited by ideologists, that examination failure at the outset of school education is widespread and that it mainly affects children from disadvantaged social backgrounds. Since the acquisition of the written language is the main concern of the school at that stage, the teaching of reading skills is then called into question; the failure to learn to read appears as the first examination failure and the one most pregnant with consequences.

More recently, the widespread desire to revitalise the teaching of reading skills on the basis of the findings of the language sciences has opened up new vistas. Linguistics, and more precisely phonetics, was the first to make a breakthrough which was fairly well received. On the other hand, later pronouncements based on psycho-linguistics have aroused violent polemics.

These two approaches, the socio-ideological and the technical-linguistic, correspond to different areas of knowledge and/or action. They are consequently independent of each other but, in some cases, they meet and become intertwined in a complex set of problems. We will now take a closer look at what the discussions are about.

1.3.3 The socio-ideological approach

In the case of the first approach, the sociological factor is not really in dispute. Once it is admitted that poor readers are basically recruited from disadvantaged environments, a problem arises for the teachers: they live with a bad conscience and, at the same time, experience a feeling of helplessness.

The most common response when confronted with this situation which has to be acknowledged but to which people do not know how to react, is to transfer the responsibility for it outside the school. The ideological interpretation of the socio-cultural handicap is therefore to attribute to the social and family environment the main responsibility for the difficulties of children when learning how to use the written language.

In the case of reading and writing skills and therefore of the language, the socio-cultural handicap is treated as a socio-linguistic handicap; the children from disadvantaged backgrounds have difficulty in learning to read because the spoken language in the home is imperfect. Some writers however are not satisfied with that explanation and, on the principle that "examination failure is not fatal" (CRESAS, 1981), question the comforting concept of the socio-cultural handicap (CRESAS, 1978) and, more specifically, even the notion of dyslexia (CRESAS, 1972). They produce evidence that through activity in the school, even the most disadvantaged children can learn to read and write without problems (Chauveau, Rogovas-Chauveau, Leuleu and Galeyrand, 1985).

1.3.4 The technical-linguistic approach

In the case of the teaching of the written language, seen from the more strictly technical viewpoint we have mentioned, there are three schools of thought opposed to each other.

The first school of thought maintains that learning to read requires the child's initiation to the grapho-phonetic code at the outset. The reference to phonetics when teaching the written language arises in this context. Without disputing the traditional approach that learning means the acquisition of grapho-phonetic relationships applicable to the French language, phonetics has enabled that approach to be brought up to date. (Huot, 1981; Inizan, 1978). But the teacher in the classroom has difficulty in regarding this reference to phonetics as anything more than "cosmetic" where the most traditional concepts are concerned. The importance attached to phonetics has strengthened the faith of primary school teachers in a simple, basic model of the relationship between the spoken and written language, but the numerous "exceptions" they encounter disconcerts them. They teach as if they were embarrassed by the French language!

Insights from the psycho-linguistics of reading skills have given rise to a school of thought which considers the meaning as more important than the code, and consequently favours direct access to the meaning through "ideo-visual" reading (Foucambert, 1976). The notion of anticipation plays a major role in this process. The emphasis moves from the language to the reader, from the characteristics of the language to the activity of reading. The provocative assertions of the exponents of this new radicalism have the no small merit of having effectively challenged the preconceived ideas regarding the teaching of reading skills.

The third school of thought tries to combine contributions from the previous approaches (Lentin, 1977; Downing and Fijalkow, 1984). It retains the second school of thought principle that the meaning must be sought without delay, but considers nevertheless that an explicit knowledge of the code is one of the ways of reaching the objective. This is a pluralist approach resulting from a less radical interpretation of psycho-linguistics; it rejects the notion of reading as decoding but keeps the requirement for special attention to be paid to the code. This approach, with the code now subordinated to the reconstruction of the meaning, will satisfy many teachers who, while accepting the basically meaningful nature of the action of reading, are not prepared to deprive the child learning to read of resorting to the spoken language.

Consequently, the debate in its usual form is a confrontation between the exponents of deciphering, of comprehension, and of comprehension partly helped by deciphering, but it also arises in connection with more precise practical points. For example, the question of reading aloud and silent reading is the subject of particularly lively controversy. The traditional view is that reading aloud is what reading is about. The defenders of ideo-visual reading, therefore, vigorously reject it. For the third school, it is legitimate if there is a communication situation. Examples of such controversies could be multiplied, but it might perhaps be better not to prolong our consideration of these debates and to look at the question from a different viewpoint.

We should be aware of the fact that the debates for most part take place among the teacher trainers, that is to say among people with a background in the teacher training establishments (inspectors, college of education lecturers, educational advisers, etc). Researchers in their university laboratories approach these matters somewhat differently, and the teachers in the schools in yet another way.

1.3.5 What really happens in classrooms?

Once the theoretical approaches have been determined and relativised, the next step is to ascertain who, effectively, the audience is in educational practice. The question is, in fact, what really happens in the classroom? How are the children in the schools actually taught to read and to write? No-one is in a position to reply objectively to this question, because no descriptive material of educational practice exists where writing is concerned. In the absence of research specifically dealing with the matter, we can merely offer a few brief and completely personal comments.

One piece of evidence is that the "reading methods" traditionally included in teaching textbooks do not exist in practice. Almost all teachers in fact use what is called a "mixed" approach consisting of beginning with a so-called "global" stage and moving on very rapidly to a systematic study of the relationship between sound and symbol. While the selection of a "method" does not, therefore, constitute a means of differentiating between teachers, the use or non-use of a reading textbook for all the class can be seen as more informative.

There is little doubt that the majority of teachers still use a textbook. One survey indicates that among the most frequently used teaching tools the textbook comes third, after the blackboard and exercise sheets prepared by the teacher (Ministere de l'Education, 1980). This information suggests that the teacher may be using a textbook most of the time, but is not content with following it and supplements it with self-produced material.

The teaching of writing skills constitutes the grey area in the teaching of the written language. It is not a subject of debate, nor is it even a real teaching subject.

In most schools the phobia aroused by the spelling mistake, more or less equated to a mortal sin, leads to the complete repression of any writing activity at the outset of school life. The authorised written exercises consist of little more than copying down prepared examples. It is only after one or, more often, two years in school that children are allowed to use the written language as a means of communication. They then pass without transition from prohibition to prescription; between the two no teaching of writing skills has taken place. There is no methodology for the teaching of writing skills. This situation, where writing skills are concerned, arises, no doubt, from the implied principle that knowing how to read is necessary and sufficient in order to know how to write.

Very fortunately the picture above relates to most, but not to all, classrooms. Alongside the traditional France with its deeply entrenched teaching habits, a pioneering minority working in close contact with teacher trainers are experimenting with other approaches. These approaches are achieved in numerous ways; we shall limit ourselves to mentioning a few. In the case of written texts designed for children's reading development, some teachers have overcome the division between the textbook and texts prepared in class ("the natural method").

Children's literature is increasingly used at an early stage. Publishers are producing small books specially prepared for this purpose; these are written in easy language, and are short and attractive. They meet the need to bridge the gap between learning and practising reading skills and arouse the pleasure of reading before the child even knows how to read.

The desire to teach the reading of numerous and varied written texts has led to the appearance in the classroom of what are generally called "social texts", ie written texts produced outside the school for non-teaching purposes. Among these, cooking recipes have the greatest success, but it is not unusual to find food wrappings or the local newspaper in classes of children learning to read. Here we see the first effects of referring to the findings of socio-linguistics.

There has also been in the last few years the spectacular development of library and documentation centres, replacing the out-of-date school classroom libraries. More than varied resource centres, they seek to become a crossroads of activities enabling each class to emerge from its ghetto.

Reading in general is becoming more and more authentic, and in the larger classes the texts chosen are giving way to whole books, in most cases novels. Teaching resources are being renewed and up-to-date teaching kits are making their appearance. The more adventurous are using micro-computers, but the lack of teaching software designed for learning the written language does not enable much progress to be made in that direction. The teaching of writing skills remains the poor relation where the teaching of the written language is concerned; attempts to regenerate it hardly impinge on the early stages of the learning process but have a greater effect on more advanced classes with rewriting activities or the collective composition of a long work.

1.3.6 Recent research in France on factors affecting the learning process

Although there is at present hardly any objective description of the teaching methodology of reading and writing skills in France, there has been some fairly recent research shedding a little light on the factors affecting the learning process.

A large-scale survey has been conducted by the Ministry of Education. A sample of some 20,000 pupils were followed throughout their primary school education. A sub-sample of nearly 2,000 pupils were given tests in French and mathematics during their first year at school. Several publications present this survey and its results (Aubret, 1985; Blanche and Le Laidier, 1982; Levasseur and Chassaing, 1984; Ministère de l'Education, 1980, 1983; Seibel, 1984; Seibel and Levasseur, 1983).

Two other research projects (Mingat, 1984; Preteur and Fijalkow, forthcoming), both concerned with a sample of 16 first-year classes in disadvantaged environments, have also evaluated the attainments in reading skills and in mathematics in the first year.

If we follow the four variables found in the three investigations, the following conclusions emerge concerning reading skills: the social background appears in every case as an important factor of differentiation between children. Thus, for example, in the Ministry survey (Seibel, 1984), the rate for repeating the first year is 2.4% for children of parents in the higher management and professional classes and 29.99% for children of farm labourers. The three investigations also agree in observing that children of foreign families have lower crude results than those of children of French families. They differ, however, when other variables are taken

38

into account; all things being equal however, Mingat (1984) finds that the children of migrants obtain better results while Preteur and Fijalkow (forthcoming) reach the opposite conclusion. Unanimity reappears when age is considered, but in a direction contrary to preconceived ideas. It is curious to observe in fact that in none of the research do the children's results vary according to their age.

Where the sex of the child is concerned the conclusion from these investigations is less obvious. In the actual reading tests, boys and girls did not obtain different results in two of the surveys (Preteur and Fijalkow, forthcoming, Seibel, 1984), but they are different in the third where the girls obtained better results (Mingat, 1984). If we consider a more global indicator, namely the rate for repeating a year, we find that it is the same after one year in school (Blanche and Le Laidier, 1982) but after four years the girls are ahead of the boys. As is often the case in the literature on the subject, we are bound to observe that there is not always a difference between boys and girls in reading development, but where there is a difference it is in favour of the girls.

1.3.7 Explanations of reading difficulties

Now that we are a little more aware of the children for whom learning to read is a problem, we can turn our attention to the explanations proposed.

Five main approaches can be distinguished (Fijalkow, 1986): the organicist, the cognitive, the affective, the socio-cultural (handicap argument) and the pedagogical. These various approaches correspond to relatively constant epistemological and/or professional viewpoints. However, as we are now specifically concerned with the existing situation, we shall not spend any more time on these problems.

Returning to the three investigations, it will be recalled that they assess reading skills and mathematics. It is interesting, therefore, to consider the results obtained by the children in each of these disciplines. One striking fact which emerges is that the results in reading skills and in mathematics coverage. Seibel and Levasseur (1983) observe in this connection that "the majority of pupils achieved simultaneously high, medium or low results in French language and in mathematics". The same phenomenon occurs in the two other investigations. This finding seems of major importance. It means that, in addition to the specific aspects associated with learning the written language and learning mathematics, there are factors common to both learning processes, non-specific factors of such an importance than they lead children to obtain similar results whichever discipline is concerned. It also means that, as we reach the point of formulating proposals designed to improve the present situation, we must not limit them to the sphere of reading skills in the strict sense, but also take into account factors relating to school life in general which are likely to have a particular effect on reading and writing skills.

1.3.7.1 Some measures to overcome reading difficulties

A first series of measures relate to teaching practices to be introduced or developed in the classroom. Getting across the content of the message before the technique, whatever it may be, seems of major importance when dealing with young children. The mistake common to all "reading methods" is that they are precisely reading methods, that is to say devices invented for acquiring technique and not content, know-how and not knowledge.

It would seem preferable, in our view, to abandon the use of graded teaching material based on inadequate scientific principles and introduce instead authentic reading situations from the outset. That approach to the text would be motivated by the expectations of the child and not by the rationale of the teacher. The increasingly wide use of children's literature as teaching material is a step in the same direction. In these conditions, reading activity has every chance of being rapidly grasped by the child and the expenditure of energy required no longer causes problems. The separation between learning time and practice time is removed. Similarly, the pleasure of reading runs side by side with the learning process.

For reading skills to be acquired much more naturally, that is to say in a random fashion as is the case with the spoken language, full attention should be given to the selection of motivational themes for the child, to the selection of books and to activity and/or research projects. The reading of specific material (a notice, a recipe, etc) will be required for the proper completion of projects, contrary to traditional teaching methods. This "deschooling" of reading skills means that a variety of written texts can be assembled and the child can learn to read several types of reading matter differentiated according to the purpose for which the texts have been selected. The act of reading is achieved in a communication situation; it is functional and no longer arbitrary.

The length of the written texts prepared for children in such cases will hardly ever be shorter than the sentence, with the child's activity continually directed towards finding the meaning. The teacher's role is then to help the child to discover the meaning of the message with the assistance of all possible indicators for that purpose. The verbal material contained in the texts can be incorporated in reading kits and games; if used flexibly, they enable what has been learnt to be consolidated without boredom.

The teaching of writing skills must no longer be postponed indefinitely. Writing activities must be developed from the outset, and, as in the case of reading, be purposeful and functional. Once again only active use of the tool will enable its function to be understood and motivate the learning process.

This brings us to the epistemological problem of the learning of the written language. One of the difficulties encountered by the approach we are recommending is the dominant empiricist theory. This approach is influenced by structuralism, and is based on the principle that knowledge is produced by the activity of the subject and is not a passive deposit imparted by the efforts of the teachers.

It views the learning of reading and writing skills as a re-invention by the child of the system of the French written language, a re-invention to be made by each child through reading, and particularly writing, activities. The underlying concept of reading activity is also structuralist. Reading activity is viewed as the action of reconstructing by the reader the meaning encoded by the writer.

The implementation of this approach cannot be achieved in teaching of the collective type within the traditional framework of a classroom where each child is regarded as a fraction equal to all the other fractions of the unit forming the class. On the contrary, the existence of individuals and groups must be recognised in order to enable the teaching of the written language to change direction. This leads in practice to organising the classroom in a different way. Organising the class into rotating workshops where the teacher works in turn with each of the groups or individuals while the rest of the class works independently is a response to the need to develop social interaction (CRESAS, forthcoming) and individualised teaching.

This method of organising the class also has the merit of putting an end to the fussy supervision which surrounds any reading activity in class at present and allowing habits of independent work to develop very rapidly.

Another series of measures concerns the school as a whole. The development of a library-documentation centre open to all classes is an interesting step towards the destruction of the ghetto constituted by the individual class. The library-documentation centre, with its exhibitions, meetings with authors, etc, encourages working in the school outside the classroom.

Forming a team of teachers responsible for the first stage of teaching the written language seems to be essential in order to ensure teaching continuity, and also to give responsibility to a group rather than a person.

This stage can be planned to cover a period of three years, beginning with the last year of nursery school and continuing during the first two years of primary school. One teacher is made responsible for the latter two years and consequently works with the same group of pupils during that time.

To enable the team of teachers for the first stage to carry out this collective operation satisfactorily, they are allowed a half-day for planning each week in the school timetable.

It would appear, subject to confirmation by a study of intra-school mobility of teachers, that the teachers in charge of the first year are sometimes teachers who have remained at this level and achieve good results, but more frequently they are the least experienced or the most recently appointed teachers. The latter category is fraught with examination failures and calls for the introduction of a policy designed to put an end to the practice.

In addition to the teaching team to be formed for the reading foundation stage, there is the question of the composition of the school teaching body itself. The existing method of constituting the teaching body is based exclusively on the individual. It is not surprising that such recruitment methods make it extremely unlikely that homogeneous teaching teams will be formed. It would seem desirable, therefore, if such teams are to exist, that an interest in involvement in a teaching project should be another factor to be taken into account when an appointment is made.

Encouraging every school to have a teaching project prepared by teachers willing to work together in a structured way is, in our view, a reasonable method of preventing innovators from becoming sooner or later completely frustrated and, in general, of allowing forces latent in the present fragmentary state of the education system to come to the surface.

Three conditions must be fulfilled for the effective functioning of these teams:

- the involvement of the head of the school who, having master-minded the teaching project, will become its main leader;

- the development of in-service training where the working unit is no longer the individual but the teaching team or part of the team concerned with a particular project. Thought can be given to the participation of a teacher trainer, particularly at the time when the first stage team is being constituted, as being the best way of meeting the real needs of the teachers;

- allowing parents, within limits to be determined, to register their children in a school where they are in support of a particular teaching project.

At the end of the teaching project, planned for a fixed period, a report is prepared which includes standard items asked for by the administration to supply some of the information normally produced by the inspectorate.

The third requirement is to re-situate the school in its environment.

The path diffidently marked out by library-documentation centres and micro-computers must be widened. Since this equipment is now obtainable in the school neighbourhood by parents in particular, we know that in this case legislation precedes demand and only in rare cases has school supply been followed by substantial results.

The school far too often remains the property of the teachers, who are as hesitant about opening up the school as adults are about entering it.

Among the measures to be envisaged for removing this barrier, two appear simple and are without financial implications:

- arranging for parents and parents' associations to participate in the management of library-documentation centres under the responsibility of the teachers;

- organising seminars of the university summer school type where, instead of the usual pattern of bringing together as individuals people who will not see each other again, the various partners (teachers, librarians, social workers, local elected representatives) who will be working together in a particular neighbourhood will meet in order to collaborate in the preparation of their action project. The idea of creating a local network seems promising in this context.

It would seem that some of the present difficulties could be reduced by a number of measures adopted in the classroom itself as well as in the school and in the school environment. The measures we have recommended require increased respect for individual differences as well as greater possibilities for team work.

BIBLIOGRAPHY

Aubret J (1985) Etude de quelques aspects de l'hétérogenéité des élèves en français à l'issue du Cours Préparatoire, *Enfance*, 4, 367-387.

Blanche M and Le Laidier S (1982) Le devenir en 1979-80 des élèves scolarisés en CP en 1978-79. Analyse des correspondances, *Education et Formation*, 1, 17-42.

Chauveau G, Rogovas-Chauveau E, Leuleu M and Galeyrand G (1985) Les ZEP mode d'emploi, *Collection Cresas*, No 4, 151-178.

Cresas (1972)*La dyslexie en question*, Paris, Colin.

- Le handicap socio-culturel en question, Paris, ESF, 1978 - *L'echec scolaire n'est pas une fatalité*, Paris, ESF.

- *Les interactions sociales à la crèche, á l'école maternelle et à l'école primaire* (forthcoming).

Downing J and Fijalkow J (1984) *Lire et raisonner*, Toulouse, Private.

Fijalkow J (1986) *Mauvais lecteurs, pourquoi?* Paris, PUF.

Foucambert J (1976) *La manière d'être lecteur*, Paris, Hatier.

Huot H (1981) *Enseignement du français et linguistique*, Paris, Colin.

Inizan A (1978) *27 phrases pour apprendre à lire*, Paris, Colin.

Lentin L (1977) *Du parler au lire*, Paris, ESF.

Levasseur J and Chassaing F (1984) Evaluation de l'enseignement à l'école elementaire, *L'Orientation Scolaire et Professionnelle*, 13, 1, 5-16.

Mingat A (1984) Les acquisitions scolaires de l'élève au Cours Préparatoire, *Revue Française de Pédagogie*, 69, 49-64.

Ministère de l'Education (1983) Evaluation de l'enseignement à l'école primaire, Year 1979, Cycle Préparatoire, *Etudes et Documents*, 1980, 3 *Note d'information*, No. 82-28, 29 August.

Preteur Y and Fijalkow J (forthcoming) Etude differentielle de l'acquisition de la lecture et des mathématiques au Cours Préparatoire, *Revue Française de Pédagogie*.

Seibel C (1984) Genèses et consequences de l'echec scolaire: vers une politique de prévention, *Revue Française de Pédagogie*, 67, 7-28.

Seibel, C and Levasseur J (1983) Les apprentissages instrumentaux et le passage du Cours Préparatoire au Cours Elementaire, *Education et Formation*, 2, 3-24.

1.4 READING –
A DEVELOPMENTAL SKILL?

by

Mogens JANSEN, Denmark

1.4.1 SUMMARY

This report centres on the youngest age levels in school. One consequence of a view of *reading as a developmental skill* is that reading must also (here very briefly) be seen in relation to the middle stages of the school, the young, the group of adults in general, and the aged.

In all the groups, reading is regarded as something absolute and clear — something a person is either *capable* of or *incapable* of: "Have you learned to read?" we ask the 8-year-old. "X% are unable to read," it says in an article. Perhaps such a starting point is inadequate?

In the report, reading is evaluated principally as a developmental skill. Reading is seen as something which differs characteristically from one stage, one level, one phase, etc. to the next:

Firstly, *the development of reading over two generations in Denmark is considered* [p 46].
Further, *reading is viewed very briefly in an historical perspective* [p 51].
After that *reading is seen more specifically in relation to the development of the individual;* and it is maintained that there are, in principle, three or four different reading levels [p 52],
A model outlining *the interaction between reader and text* is mentioned on p 57.
Finally, some *educational and social consequences* of the four above-mentioned starting points are described, especially in connection with the disabled readers. [p 58].
At the verbal presentation of the report, the influences of research, if any, will be discussed, and *research areas to be taken up* — preferably through a cross-national research cooperation — will be pointed out.

1.4.2 THE DEVELOPMENT OF READING DURING A COUPLE OF GENERATIONS IN DENMARK

1.4.2.1 *Reading development in children, seen over a number of years*

In Denmark it is not possible to follow reading education by means of annual and nation-wide tests. Such tests are simply not applied. Neither is material collected with a view to standardization, and thus reading development cannot be described from year to year throughout schooling in this way, either.

Nevertheless, it has been possible, with the aid of school psychologists, reading consultants, and others, to collect data which have, after statistical processing, yielded sufficient information to outline the development of the reading competence of the pupils from the years after World War II to the present. See *Note 1.*

A summary of a number of studies is indicated in *Figure 2,* cf. Soegaard, Jensen & Hansen, 1977; Jansen & Glaesel, 1977; Jansen & Kreiner, 1986; furthermore, cf. Elbro *et al,* 1981; Dalby *et al,* 1983. *Figure 1* reproduces, according to Soegaard et al (1977) the first establishment of a trend which proved to be general. Most recently the same pattern is seen in Jansen & Kreiner (1986), now showing up to 7% in the extreme groups. Figure 2 shows a stylized version of the general pattern:

1 Today the pupils learn to read later than was the case a generation or two ago.

2 There are far more *better* readers than there used to be (C); there are also some pupils whose reading is *too bad* (A). Also this group is increasing, although essentially less than group C.

3 All the pupils read more than previous generations (this is not shown in the figure).
   These features have often been established, at least since the mid sixties. During the last decade yet another dimension has come into the picture - reading ability increases "in waves".

4 Particularly in grade 2 (when the pupils are about 8) there are many weak readers. In grade 3 the number of good readers increases. In grade 4 too many weak readers appear again, but there are also many good readers. In grade 5 a clearly increasing tendency to more good readers is seen.
   It seems as if the varying increases and decreases, which are seen primarily at the outer points of the distributions, must be considered in connection with what Downing (1983) calls the phase of realization, the phase of mastery, and the phase of automation. The pupils *realize* that they must learn to read; and gradually they *master* reading at a certain level. However, due to *lack of practice* in school, not all the pupils have their reading ability automatized. Thus a growth is not seen until point A, and gradually a (larger) growth is seen at point C.

Figure 1

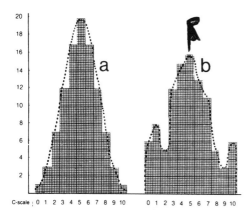

C-scale 0 1 2 3 4 5 6 7 8 9 10   0 1 2 3 4 5 6 7 8 9 10

In drawing, **a** the squared area shows the distribution of pupils classified according to reading ability after a reading test in grade 3, autumn 1966; 10 indicates the best group.

In drawing **b** the squared area shows the distribution of pupils classified according to reading ability after the same reading test in the same grade in 1972 and 1973.

The dotted lines show how the "squeeze" tendency has influenced the original distribution.

Figure 3

A changed consumption of library books:

mio./year

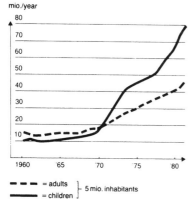

1960   '65   '70   '75   '80

— — — = adults ⎫
———— = children ⎭ 5 mio. inhabitants

The figure shows the book lending in million lendings per year from public libraries, described via lendings to adults and children, respectivly.

A fall in lendings to children after 1980 might be expected in connection with the dropping in numbers of children at the reading age, but no fall has been confirmed. The increase in lendings to adults from the same year must be seen in relation to the fact that the children in increasing numbers continue as readers into their youth and that the young ones continue reading as adults.

Figure 2

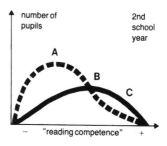

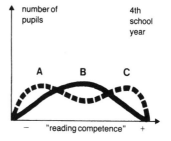

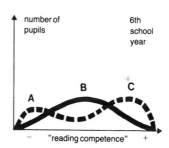

Roughly outlined the development of reading abilities among Danish children looks like this in grades 2, 4 and 6 – earlier and now.

1) generally, they read **later** (not "better" – not "worse");
2) they all read **more** (cannot be seen here);
3) there are **far more** better readers.
4) there are also some who read **too badly** (are "reading handicapped").

████████ earlier

■ ■ ■ ■ ■ ■ now

47

5 Beyond the above-mentioned points 1-4, weak but unmistakable tendencies to an *increased* squeeze effect are seen during the period from 1976 to 1984. However, according to the above, the tendencies do not penetrate consistently in all grades.

If one goes into particulars, following at the same time the development of classes and of individual pupils, the pattern outlined above is seen clearly. What Dalby *et al* called "a St.Matthew tendency" is perhaps an informative description with a clear reference to the words of the evangelist, "Unto everyone that hath shall be given...". One pupil in grade 1 read 1,270 pages in a year; another pupil read 160 pages. One pupil in grade 5 read much more than 10,000 pages in a year; another pupils in the same class read about 100 pages". (Dalby *et al,* 1983).

1.4.2.2 Why does the school fail during the automation phase?

The class quotient characteristic of the Danish school (below 19 at an average) — which is very low from an international point of view — has partly been 'paid' with a cut in the time of instruction. By way of example, the mother tongue instruction (in Danish) (i.e. "reading and literature") in the grades 2-6 has *decreased* by 34%, 37%, 45%, 47% and 27% respectively, from 1962 to 1984 (Oregaard, 1986); add to this the fact that Danish children spend a shorter period of time in school than children in other countries. During the same period both the amount of the subject matter and the requirements in general have been *increased,* as far as these grades are concerned.

This means that the pupils have had no possibility whatsoever of reading to a sufficient extent during their class hours.

The practice of reading (i.e. the work with the automation phase) has been transferred to the leisure time of the pupils and has to a very high degree been controlled by the pupil's own taste for reading — and by the library facilities available. When Danish children read largely "because they think it is fun" (Allerup, 1985; Allerup & Jansen, in process), this is extremely positive from many points of view.

However, a fact which cannot be concealed is that the pupils in group A (cf. Figure 2) have, to some degree, 'paid' for the general distinct growth in amount of reading, taste for reading, and reading competence.

1.4.2.3 Libraries, selection of books, and the use of books

If these research results are compared with changes in the lending from school libraries and children's libraries, it appears that the growth in children's lending has been constantly increasing since 1963, cf. *Figure 3* (Dalby *et al,* 1983). This should be seen in relation to *Figure 4* (Kuhl & Munk, 1979), underlining that children are mass borrowers at the library even from their very first years in school, and that many children are borrowers (and users of books) long before they start "learning to read".

Figure 4

Age grouping	So many percent of the age groups are borrowers	
	1968	1979
0-6 years	36%	37%
7-11 years	70%	95%
12-15 years	75%	98%
Adults	40%	48%

Figure 5

	1952-54	1966/67	1981/82
Easy books, fiction	148	327	932
School-books	149	243	895
Easy nonfiction	0	241	1.357
	297	911	3.184

Figure 6

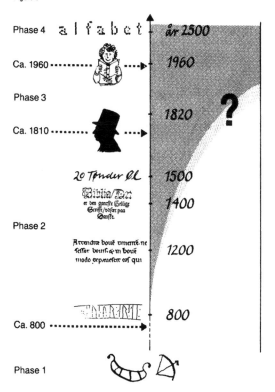

49

The amount of children's books is largely proportionate to the size of the language area. Since 1977, about 1,000 children's books have appeared yearly — in the last decade as many books as appeared totally from 1900 to 1970 (Weinreich, 1986). Perhaps the most remarkable fact is that not only the reading of fiction, but also of books supporting the pupils' personal fields of interest comes across strongly, even more strongly than their use of books in connection with the school. To a larger degree the children borrow books from the school library in order to use them at home (Jensen, 1985). The same experiences are described in Jensen (1984a; 1984b; 1985c).

Again, books of *non-fiction* borrowed and read by the pupils are used more at home than in school. Even many reference books are borrowed and taken home: more than 1/3 of the reference books are borrowed by the pupils to be used out of school for their own entertainment. This is also true of the easily accessible books of non-fiction which amount to more than 1/3 of the reading matter of the children (Jansen, 1985).

The background to these facts is that since the early fifties the book collection of the libraries has been changed from being more or less directed at the best readers (the mass readers, the elite readers) to including books for persons at all reading levels. This extension also means that now a broad range of pupils is able to *utilize* both the school library and the public library to a considerably larger extent than before.

Jansen (1983) carries an account of the growth in *easily readable books* at the Danish school libraries and children's libraries: in 1952-54 only the following amounts of books were available: "148 easy books (of fiction) for the children's leisure reading", "149 relatively easy books for use in school", and "O easy books of non-fiction". *Figure 5* shows how especially the easy books of non-fiction have increased. One of the results has been that, not least, quite a lot of the boys have become interested readers, exactly by using books of this genre. A major background is a strong growth in the drama- documentary book (Jensen, 1985). This is specifically described in Jansen (1986), and is a type of book used for the voluntary leisure reading of the pupils down to the age of 8-9. This type of book has become essential in Denmark.

Over a period of about 40 years, children and young people have thus become experienced users of books to a degree which would have been unthinkable a couple of generations ago.

Further, it gives food for thought that both in relation to the public library system and to the books they are reading or using at a given time, children and young people show the same *consumer pattern* as highly qualified adults: a very large part of the groups of pupils use the library functionally and most objectively. Not only fiction is of interest to them, but decidedly non-fiction as well. The reference books of the library are also used by these groups.

50

1.4.2.4 The sales of books seen over a number of years

Total book sales in Denmark amount to 1½-2 thousand million Danish kroner. Corrected for inflation, VAT, etc. book sales increased between World War II and about 1968. Then sales became almost stable up to about 1980, when a limited decrease could be observed. (Kustrup, 1985). During recent years the sales of school books and children's books have dropped drastically, but so far this drop has only been evident through libraries being less well-equipped than previously.

However, it is more than a curiosity that financial cuts have reduced the number of books in public libraries so strongly that, for the first time since 1920, there has been a distinct decline in book lending — due to the fact that adults have deserted the library.

1.4.2.5 Particulars about the use of books by disabled readers

It is encouraging that, on average, pupils in special education (by far the largest group among these pupils in Denmark are disabled readers) borrow more books than other pupils every time they visit the library. Naturally, their books are often smaller, but then disabled readers take a longer time to read these books.

Even more encouraging is the fact that the increased book lending to disabled readers is not for direct use in special education. Also, these pupils have an increased lending of books "for themselves" and "for the subjects in school" — this is an often ignored, but very essential result of the effects of special education in reading. "A large part of the methodology of special education aims at improving the reading ability of the pupils by stimulating to more leisure reading" (Hors and Jensen, 1983). In other words, the methodology seems to have been successful.

1.4.3 READING DEVELOPMENT IN AN HISTORICAL PERSPECTIVE

1.4.3.1 From Runes Carved in Stone to Light on Screens

Figure 6 indicates reading development in an historical perspective. Ahead of what the figure shows is a much larger historical perspective, cf. Schmandt-Besserat (1984), making it probable how writing and reading could have come into being.

What is exciting in Figure 6 in this connection is the very way in which writing was established: writing tools and materials have been essential to reading. It is important to recognise that in parts of the world where information technology is highly developed, we are on our way to a new technique of writing and of reading.

The first phase indicated in the figure (1) shows pictures — the abstraction towards syllables or letters has not yet taken place. As the writing (2) is carved in stone or written on parchment or paper, more room is left at each change of material for more writing, more information, more varied texts, etc. "With Gutenberg's movable types a monopoly of information and knowledge was broken" (Roberts, 1982). Now the line towards the 19th century, and the era we are now about to leave, is outlined clearly.

We are about to take the next step — with writing drawn by light on a screen. The technique calls for competence, and offers new possibilities. This will become important for the schools, and for beginning instruction (Lundahl, 1985; 1986) — but that is another story.

1.4.4 READING SEEN AS DEVELOPMENTAL PROCESS

1.4.1.1 Developmental psychology has inspired reading researchers

If the reading development of the individual is seen in relation to the view of development taken by a number of researchers, much general research within developmental psychology leaps to the eye — Bruner, Piaget, etc. By way of example, this applies to the clarification that the intellectual development of children takes place in certain stages; the sensorimotor stage, the pre-operational stage, the concrete operational stage, and so on.

Again, reading researchers have shown interest in phased development within this special area, e.g. Chall (1983). From the North, Sundblad, (1981) and Lundberg, (1986) can be mentioned. When researchers such as these have been especially inspiring in Denmark, the reason is, perhaps, that it has been possible to build upon a tradition which, in principle, views reading as a developmental skill and describes reading in relation to reading consumption as well as to the development of reading materials. Jansen, (1959), Jansen *et al*, (1978).

This view of reading is perhaps reflected in the structure of the Danish educational system. At any rate, it has furthered *a developmental view* of reading.

The teacher who starts instructing the 6-7-year-olds in reading continues to teach them in reading and literature, at least until they are 11-12, and in most cases until they are 16-17. This means that the teacher is spontaneously confronted with the development of the individual pupil, whether it goes smoothly or in jumps. The teacher cannot fail to notice how at some times the pupil absorbs great quantities of literature of a certain genre, a certain degree of difficulty, etc., and at other times seems to make no progress — although perhaps both teacher and pupil have a clear impression that the pupil is on his/her way towards the next reading level.

The increasing demands of the information society on the reading competence of the population, and thus on the teaching of reading, are not directed exclusively to out-of-school education and adult education, or to the oldest grades in school. At least from the age of 10-11, it is necessary that reading instruction is 'aware' of the strongly utility-oriented demands made on reading by the information society.

There is no discrepancy between a literary angle of approach to the teaching of reading and a utility- and communication-oriented angle of approach. Both approaches are necessary in order that the pupil is able to function in a society which builds on know-how and the exchange of information.

Without competence in the reading of subject material it is impossible to build up an educational system capable of meeting the demands of the information society - and unless at the same time the creative and imaginative aspects of reading instruction are considered and encouraged, primarily through the reading of fiction, the personal growth of the individual will be weakened. And so, too, the reading of subject material will have poorer conditions.

1.4.4.2 Three (four) different groups of readers

Reading as seen in the light of the reading development of the individual falls into three different groups: *A — the reading of the rebus reader, B — the reading of the transition reader,* and *C — the reading of the content reader.* And before it can even be called *reading,* a *reading of symbols* is effected, cf. *Figure 7.*

We assume that the child is a *rebus reader* from the age of about 6 to about 8. At these levels we are perhaps one to two years behind, for example, England and the United States. During this period the child learns a few relatively easy, maybe characteristic words. The child becomes familiar with letters — he/she is able to connect letters with sound. Largely, the child learns what in the old days, 150 years ago, was called "reading".

Today the child of 7-9 can read very simple texts which he/she often sees, or which are easy. In the old days the child could read the Lord's Prayer, the Ten Commandments, most well-known hymns etc. In countries where political manifestos are learnt by heart, children at this age can learn these as well.

The transition reader is between 9 and 11. But there are transition readers who are only 8, and there are children who do not become transition readers until the age of 10-11. The overlap between different reading ages and actual ages is quite considerable.

The child is now able to read reasonably accessible sentences. Naturally, he/she is also familiar with easy as well as somewhat more difficult words, and possesses supreme skill when it comes to letter recognition, and so on. Moreover, the transition reader can read easy texts. Fiction seems to be easier to read than non-fiction. Perhaps this is due to the fact that the concepts are easier; at least they are often more well-known than the concepts in non-fiction.

It seems that at the age of 9-10-11 there is a developmental barrier for children to

Figure 7

Reading development: ▶	Symbol-reader	A rebusreader 1st-2nd (3rd) grade	B 'transitionreader' (2nd)-3rd-(4th) grade	C contentreader Abt. 4th grade and up

Figure 8

Reading development: ▶ Evaluation criterion ▼	Symbol-reader	A rebusreader 1st-2nd (3rd) grade	B 'transitionreader' (2nd)-3rd-(4th) grade	C contentreader Abt. 4th grade and up
I The personal interest of the individual in the contents of the text		1	2	③
II The linguistic form of the text		4	⑤	6
III The immediate visual appearance of the text		⑦	8	9

Figure 9

Figure 10

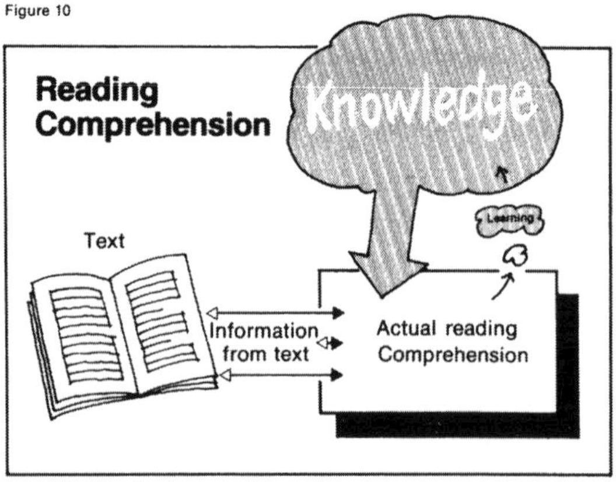

Reading Comprehension

Knowledge

Text

Information from text

Learning

Actual reading Comprehension

cross. To use a catch phrase, one could say that during the years of transition reading the child must go from 'learning-to-read' to 'reading-to-learn'. This is consequential in connection with the material we give to the children. We have not learnt how to handle this.

At the middle stages, where we have the transition readers, there are many children who *have* learnt to read. But we have not taught them how to *use* what they read.

Usually *the content reader* does not appear until about the age of 11. But, of course, we do have content readers at the age of 9-10. Gradually the content reader reads even difficult texts with a good comprehension. He or she is fully familiar with relatively easy texts. The content reader knows the common linguistic terms, set phrases, etc. and knows what can be skipped in books, newspapers, magazines, etc. with relatively stereotyped subject matter.

There are *quantitative differences* between the reading of the three groups of readers. The amount they can cope with differs greatly. It is, however, far more essential that there are also *qualitative differences* between the reading of these three groups. This means that we must also be aware of the differences in comprehension between the three groups of readers.

A great task is to adapt the educational material both to the individually and to the socially attuned reading level of the pupils. *At the same time,* we need to develop a methodology where the reading techniques based on practice are subordinated to the "whole" reading methods resting on comprehension.

This comprehension must then be constantly supported by reading techniques enabling the pupils to comprehend still larger parts of texts and still larger vocabularies and sentences. The subordinate whole must be maintained, and at the same time it must be varied and specified in the subsequent practice.

This is the demand made on reading instruction which has to be *changed* concurrently with the development of the pupils.

Almost provocatively briefly, it could be put this way (Schmidt & Schmidt, 1984):

> "*The rebus readers* learn to read what is said *on the lines.* Gradually "they can read", no doubt about that. *The transition readers* learn to read what is said *between the lines.* They are now able to draw the concepts from the texts. They can read in order to see opinions somewhat other than those apparently suggested by the texts. They can manage somewhat more difficult linguistic constructions, where one element draws meaning quite distant from its own place in the sentence. *The content reader* reads *behind* the lines or *beyond* the lines. It is the content reader who can cope with irony, double meanings, etc. The content reader manages to read eventful literature as well as philosophically rather advanced literature. By way of example, the content reader is able

to read lyrical poetry and to grasp the meaning of the many levels in lyrical poetry.

One has to be familiar with what it says not only on the lines, but also between the lines, and in most cases behind the lines. With these considerations we have come close to something very central in the democracy; however, this point may be far too serious!

1.4.4.3 The three parts of the text

In this presentation, reading is stressed as a developmental process.

It is suggested that reading instruction must respect, and build on, factual developmental stages.

In order to take account of this, a reading methodology must consider the various facets of the text which, from an educational point of view, can be divided into three, as follows: (a) *The contents of the book,* (b) *the language of the book,* and (c) *the visual appearance of the book.* (cf *Figure 8*).

(a) *The contents of the book* are generally more interesting to the mother tongue teacher than to the reading teacher. Personally, I think it inappropriate that reading is divided into reading instruction and literature instruction — but that is a different matter.

(b) *The language of the book* is the second part of the book. We can describe language. We can try to 'measure' it.
Some would say: "But what about the *concepts?* Do I place the concepts under (a) - the contents - or under (b) - the language?" The answer is "Yes"! The two things belong much more together than we usually assume in our daily education.
However, when talking about the language, we must bear in mind that the *language contains and colours the concepts* and that *the concepts carry and adjust the language.*

(c) The third part is *the visual language, the visual appearance,* of the book. This is not a joke: if a book looks like the latest large handbook of education, the exterior of the book tells us something about its contents and its level. Nobody would expect it to be a collection of poems. If we compare "the visible language" of "Frankfurter Allgemeine" with the visible language of "Bild", it is evident that the very visual appearance of the text is essential to the readability.

Many things other than the typography form part of this visual language, e.g. paper, quality of print, colours, choice and composition of illustrations — in other words, the design of the book, the newspaper, etc.

56

1.4.4.4 Reading comes into being by an interaction between reader and text

Now the interesting element occurs when we look at the *interaction* between the factors in Figure 8.

It would be more than a mission in life to try to connect the qualifications of the readers with the qualifications of the texts. Both are subject to constant change.

The contents of the text are practically 100% decisive to the reader's choice of texts and to the reading profit of these. Naturally, the contents are *never* immaterial — irrespective of the age of the pupil. A text without contents is nonsense. It is a waste of time — or worse. However, many discussions about the contents of elementary texts have been conducted much more on the premises of the adults than on the premises of the early readers. The qualifications of the adult enable the adult to read into the text a meaning which is different from that of the child. Nor is the visual appearance of the text and the language immaterial at any level, and at the initial level of reading, the level of rebus reading, the visual appearance of the text is essential.

In connection with teaching reading comprehension, reading with the purpose of learning and experiencing, it is necessary to be aware of the middle point, (b) - the language and concepts of the text.

When discussing *comprehension* and *the acquisition of concepts,* it is important to be particularly aware of the fact that our schoolbooks — that is, the language of the reading materials as such — seem to present great difficulties, especially to the transition readers.

The assumed background is that the rebus readers have gradually "learnt to read". They master some techniques. Now they are going to read in order to learn — and this causes trouble. Reading small words and simple words does not cause problems; neither does the connection of letters with sounds. Larger connections of words and technical terms may cause problems. And certainly the meaning behind the words does present problems.

The connection of reading

(i)   as *an individual developing skill* (where, for the sake of clearness, reading is structuralized in three groups)
(ii)   with *reading matter,* books, magazines, television subtitles (which can be divided into three groups)
has produced this field with nine spaces, and even a field in front to mark symbol readers. It is an educational experience that in practice field 7 often presents a block to rebus readers. If they have difficulties with b, d and p, possible problems at other levels are of secondary importance!

Unfortunately, it is also a general experience that field 5 often gives transition readers great problems. Not only 9-11-year-olds are transition readers. In Denmark we calculate that *about* 20% of the population are transition readers, when it comes to common texts meant for the population in general.

And naturally *3* is the big problem in connection with the meeting between the competent reader and a text which is inappropriate, as far as the contents are concerned. However, in all events the contents are essential.

However, behind the *total* field we find the individual's personal and total background of reading. This is something which includes, among other things, motivation, environment and home background, as well as physical perceptual, psychological and physiological background.

Perhaps the Danish term "laesningens okologi" can be translater into English as "the ecology of reading"? The word "ecology" has different values. I mean that the *total reading background* of the individual is decisive to reading comprehension.

Reading comprehension is an active process

A person does not read a text passively just by having it placed in front of him. The individual reads 'with himself' and thus interprets parts of 'himself' (his general and specific qualifications) into the texts read.

Two persons read the same text and both find that they "comprehend it correctly", but their intellectual and emotional background, etc. mean that, although it is "objectively seen as the same text", the texts they experience are different, cf. *Figure 9.*

Reading comprehension is not a question of a text reaching the person who then comprehends it. The reader is in constant interaction with the text — drawing from it all that he is capable of, comprehending it in his own way on the basis of his knowledge, his emotions, etc. and treating it further. *Figure 10* attempts to describe reading comprehension as an active process in which reader and text interact.

1.4.5 THE TEACHING OF READING – SOME CONSEQUENCES?

From a reading teacher's point of view, the most difficult task during the next generation will not be teaching Danish pupils reading. The methodology behind the "as-well-as" tradition of the Danish reading instruction (Jansen, 1978) is so flexible and well-functioning that it will probably be possible to adjust the methodology to changing demands. Again, reading materials seem to be constantly adjusted. Since the holistic tradition is very strong, a decline towards tendencies to the (inappropriate) sub-skill-accentuating, detail-oriented teaching of reading is hardly a threat.

Much sooner, what might cause problems seems to be financial and organizational circumstances — both in ordinary education and in the education of children, the young, and adults who have special difficulties in "learning to read". See *Note 2*.

Schools are under pressure, and during the next decade they will be pressurized to an extent which might make teachers, administrators, and educational politicians choose other professions, given the great demands made on schools, teachers, and pupils.

Education — not only during the years of childhood, but during the whole of life — will become essential. In adult education the three "R"s will become still more essential. *At least, reading will!* Parts of mathematics seem to be on their way out, but, indeed, other parts are on their way in. Writing is changing nowadays, and will probably become less important - to raise a point which is still somewhat controversial. See *Note 3*.

Reading instruction will break down unless it improves; becomes more appropriate; builds on more emotionally positive elements. And unless more importance is attached to it — that is, at the level where children and young people must *read to learn.* What is needed is reading instruction different from that which is general in the early years.

The natural scientific tradition of measuring the measurable things and making the immeasurable things measurable is not to be contested here for one second. Without that tradition we would probably not have a modern society.

Problems arise when the measurable things are actually rather uninteresting and may even obtain their essential value solely because of the fact that they can be measured.

It would be desirable to be able to measure many of the immeasurable things in reading — e.g. *the joy* of the individual child on reading "Pippi Longstocking", *the fascination* of the individual adult on not only seeing Shakespeare at the theatre or on TV, but on *reading,* for the first time, King Lear. When we are able to measure such things, we are able to measure the essential things in reading.

When we are able to measure the comprehension of a text — and more than details which can be controlled immediately — we have arrived at something vital. It is valuable enough to be able to check how many arrows Robin Hood had in his quiver at the shooting match on Finsbury Heath. And it is not unimportant that the child reading "Robin Hood" has understood the play between Robin and his outlaws, the sheriff and his men — and the King and Queen. It is important that the mind of the reader has been opened to new impressions — from the adventures conjured up by the reading of the shooting match on Finsbury Heath.

Note 1

The most widely applied reading tests so far can be described as follows: *OS400.* Group test. Grades 1-3, silent reading test of words (Soegaard *et al*, 1974) shows how many words children can read correctly within 10 or 15 minutes. The test is used for evaluating children's abilities in reading single words and relating them to a drawing illustrating the meaning of the word. — *S50.* Group test. From the middle of grade 2 to the end of grade 5, silent reading test of sentences (Soegaard *et al*, 1972). The test describes both the class and the individual pupil, as far as reading speed and reading quality are concerned. Quite a few other tests have also been applied.

Note 2 Disabled readers, a group which, relatively, is growing

In this report the disabled readers are hardly mentioned. This is not because we do not have disabled readers in Denmark! As mentioned before, this group has, on the contrary, been increasing slightly during the last decade - perhaps even since the early sixties.

The fact that the demands of society, and thus the demands on schools, are increasing generally, means that, relatively, the reading competence of disabled readers is falling.

A reading handicap is a serious problem to the individual afflicted with it. It is also a problem which is going to be seriously injurious to quite a large part of the population; we are on our way to an information society where a considerable reading competence is required.

And the democracy as such is in danger of coming to pieces, if a larger part of the population of society is 'tied off'. The basis of this is probably laid down long before the beginning of schooling, and even at the age of 8 the problem begins to make itself felt

Note 3 And what about writing?

If one is not familiar with Danish educational tradition, one wonders that there is so much talk about reading and so little about writing.

For many years there was a tradition to let reading be concurrent with writing. As the demands on reading have increased (during the last 30-40 years) there has been a tendency to rate the teaching of reading higher, without rating the mother tongue instruction in Danish lower; the latter continues at the same level.

The measurable results behind "writing" instruction can be observed in handwriting, spelling, and in written composition, whether this is an essay or about subject matter.

There are no authorized evaluations of handwriting; but it is indisputable that as the amount of what is written has increased, the quality has decreased. The transition from pen, ink and pencil, firstly to propelling pencil and then to ballpoint pen, later to felt pen and typewriter, has continued — now to the use of a keyboard for screens. We have yet hardly taken in the educational consequences of the changing materials.

Within spelling instruction a "St Matthew" tendency can be observed, as within reading, but in mother tongue instruction in Danish, development is 1-1½ years *behind* development in reading.

There are no certain evaluations of instruction in essay writing, writing about subject matter, non-fiction writing, etc.; however, at least the Ministry of Education, which is otherwise extremely critical, sees no fall in competence.

REFERENCES

Allerup P (1985) *Why I like to read – statistical analysis of questionnaire data.* Kobenhavn: Danmarks paedagogiske Institut, 61.

Allerup P and Jansen M (in press) *Why I like to read - educational analysis of questionnaire data.* Kobenhavn: Danmarks paedagogiske Institut.

Chall J S (1983) *Learning to read: the great debate.* 2 rev.ed. Introduction to the second edition. An update. New York, N.Y.: McGraw Hill Inc.

Dalby M A, Elbro C, Jansen M, Krogh T and Christensen W P (1983) *Bogen om laesning I — forudsaetninger og status* (Book of Reading I — Background, Conditions and Status). Kobenhavn: Munksgaard, Danmarks paedagogiske Institut.

Downing J (1983) *Hvad betyder skolen for barnets laese — og skrive-udvikling?* (New Perspectives on Reading and Writing) Kobenhavn: Landsforeningen af Laesepaedagoger, Laeserapport 8.

Elbro C, Jansen M and Lob H (1981) *Hvor godt laeser og staver born i dag?* (The Reading and Spelling of Children Today) Kobenhavn: Det Centrale Uddannelsesrad. Modersmalsopgaven. Arbejdspapir no.3.

Hors S and Jensen P E (1983) *Hvilke boger lanes?* — En undersogelse af, hvad elever laner i skolebiblioteket (What Books are Borrowed? — A Study showing what Pupils borrow from the School Library). Herlev: Herlev kommune.

Jansen M (1985) *Special — og begynderundervisningen i dansk.* Metodisk vejledning (Special Education and Beginning Education in Danish. Methodological Guide). Kobenhavn: Laesepaedagogen.

Jansen M (1986 in press) *A little about language, words, and concepts — or what may happen when children learn to read.* Kobenhavn.

Jansen M and Glaesel B (1977) Hvordan laeser og staver eleverne? — en sammenlignende status (The Reading and Spelling Competence of Pupils — A Comparative Study). *Laesepaedagogen, 25* (4), 193-204.

Jansen M, Jacobsen B & Jensen P E (1978) *The teaching of reading — without really any method.* An analysis of reading instruction in Denmark. Kobenhavn: Munksgaard. New Jersey: Humanities Press Inc.

Jansen M and Kreiner S (1986) *Hvor meget darligere eller bedre laeser born i dag? — en opgorelse over laeseniveauet i 2-5 klasser for arene 1976-84, sammenstillet med tidligere ar* (How much worse or better is the Reading of Children Today? — An Account of the Reading Level of Grade 2-5 Pupils for the Years 1976-84, as compared with Previous Years). Kobenhavn: Den gule serie. Paedagogiske forskningsrapporter nr 36.

Jensen H (red) (1985) *Elever og laereres faktiske udnyttelse af skolebiblioteket* (The actual Utilization of the School Library by Pupils and Teachers) Kobenhavn: Danmarks Laererhojskole.

Jensen P E (1984) Fagorienteret bibliotekskundskab (How to use the Library — Especially the Non-Fiction Book). *Born & Boger 37* (5) 4-9 (c).

Jensen P E (1984) Skolebibliotekerne — en succes, der kan gores bedre (School Libraries and Natural and Social Science) *Born & Boger, 37* (5) 4-9 (c).

Jensen P E (1985) Laererne skal kende skolebibliotekets muligheder (Teachers must know the Possibilities of the Library) *Laesepaedagogen, 33* (8), 314-319.

Krustrup E V (1985) Bogomsaetningen pa vej op (Book Sales are increasing). *Det danske bogmarked,* 45, 1383-1389.

Kuhl P-H and Munk J K (1979) *Folkebiblioteket og befolkningen* (The Public Library and the Population). Kobenhavn: Socialforskningsinstituttet Publikation 89.

Lundahl F (red) (1985) Laesning i teknologisamfundet 1 (Reading in the Technological Society 1). *Laesepaedagogen, 33* (6).

Lundahl F (red) (1986) Laesning i teknologisamfundet 2 (Reading in the Technological Society 2). *Laesepaedagogen, 34* (2), saertryk.

Lundberg I (1984) *Sprak och lasning* (Language and Reading) Stockholm: Liber.

Roberts J M (1982) *En Verden: Skaberen Europa (ca. 1800-ca. 1900)* (One World: Europe, the Creator (abt 1800-abt. 1900). Forums verdenshistorie. Kobenhavn: Forum.

Schmandt-Besserat D (1984) Before Numerals. *Visible Language,* XVIII(1) 48-60.

Skov P (1986) Skolens indhold og vilkar (The Contents and Conditions of the School). *Danmarks skolelederforening,* nr 2, Temahaefte,

Sundblad B, Dominkovic K & Allard B (1981) *LUS — en bok om Lasutveckling* (LUS — a Book about

Reading Development). Stockholm, Liber.

Soegard A, Jensen B and Hansen J (1977) *Laeseudvikling og danskundervisning — en beskrivelse af et praktisk forsog med at sammenknytte laeseproveresultater med begynderundervisningen i dansk* (Reading Development and Instruction in Danish — a description of a practical attempt at linking reading test results with beginning instruction in Danish). Kobenhavn: 1975. 2. ajouforte udgave 1977. Den gule serie. Paedagogiske forskningsrapporter nr.5.

Soegard A and Petersen S B (1972) *Saetningsstillelaesningsproven S 50* (Silent reading test of sentences, S50). Kobenhavn: Dansk psykologisk Forlag.

Soegard A and Peterson S B (1974) *Ordstillelaesningsprove OS 400 ord* (Silent reading test of words, OS 400 words). 1. udg. Kobenhavn: Dansk psykologisk Forlag.

Weinreich T (1986) *Borns boger* (the Books of Children). Kobenhavn: Host & Son.

Oregaard T (1986) Folkeskolen netop nu! (Folkeskolen now!). *Folkeskolen,* 1986 *103* (31/32/33), 1223-1224.

1.5 THE RELATIONSHIP BETWEEN DECODING SKILLS, READING COMPREHENSION AND SPELLING SKILLS IN THE FIRST THREE YEARS OF PRIMARY SCHOOL

by

J C MOMMERS, The Netherlands

1.5.1 SUMMARY

The acquisition of reading and writing skills is central in primary school. It appears that many pupils have reading and writing difficulties. A project, called 'Prevention of Reading Difficulties' was carried out in the Netherlands between 1978 and 1985.

The strategy applied consisted of three stages: (1) A longitudinal investigation to establish the extent to which it is possible to predict reading achievement before formal reading instruction starts, (2) The development of instruments and procedures to screen the pupils who need special attention, (3) The selection or construction of adequate educational procedures and programmes to stimulate the development of these children.

Some results of the first stage of the research are presented. Four specific questions are answered: (1) Is the distinction between general and specific reading prerequisites relevant to the prediction of reading and spelling achievement? (2) To what extent are aspects of reading achievement empirically distinguishable in the first grade? (3) How strong is the influence of decoding skills, reading comprehension and spelling at the various points of measurement in the first three grades? (4) Are there influences of factors at one point of measurement on factors of a different kind at a later point of measurement?

The answers to these questions are relevant for the development of course material and the diagnosis of reading and spelling difficulties.

1.5.2 INTRODUCTION

'Prevention of Reading Difficulties' was a project carried out in the Netherlands between 1978 and 1985. This project was inspired by a similar project done in Sweden during the sixties (Malmquist & Valtin, 1974). Its intellectual basis draws largely on the current debate on how to improve schools, instruction and education as quickly as possible and with the greatest substantial, cumulative and lasting effect. The relationships between research, development and diffusion were illustrated using the output-oriented model proposed by Gideonse (1986). This model is based on the conviction that research, development, and school operations are different kinds of activities with quite different objectives and outputs. In his model Gideonse distinguishes three planes, symbolizing the three different orientations of activities: research, development and school operations.

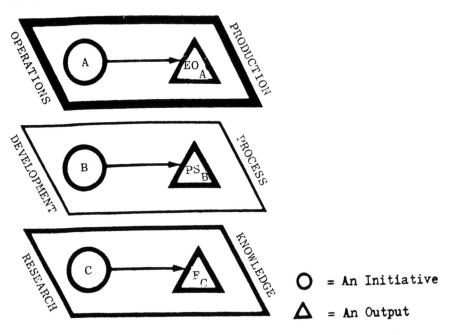

Figure 1: An output model of educational research and development (Gideonse, 1986). EO = Educational Output, PS = Performance Specification, F = Finding.

The lowest plane symbolizes the knowledge orientation of research. The object of research is to generate new knowledge. One of the significant features of research is that when an activity is begun the specific outcome is not yet known. The middle plane symbolizes the process orientation of development. The object of development is to produce materials, techniques, processes and organizational formats for instruction which accomplish certain pre-specified objectives. The

64

top plane symbolizes the activities characteristic of school operations. The object of school operations is to act upon human beings in order to train and to develop in them various skills, attitudes, beliefs and knowledge.

1.5.3 THE PROBLEM: PREVENTION OF READING DIFFICULTIES

The acquisition of reading and writing skills is central in primary school. Yet it appears that approximately a quarter of the pupils experience reading and writing difficulties in the sense that they do not achieve the expected level of reading and writing skills at a given age. School officials ask for school operations or development projects designed to remove these difficulties. A good development project will begin by looking for relevant research findings which may offer clues to guide developmental activities. It is crucial that such a search should be made at some point near the very beginning of a project. Research has shown more than adequately that reading difficulties are the result of a complex of causes: physical, intellectual, environmental and emotional. Most of the research among children with reading and writing difficulties has focused on diagnosis and remediation. Little research is known that focuses on the prevention of reading difficulties and maintains the students, as much as possible, in their normal learning group. A series of research projects from Sweden (Malmquist, 1969) is among the few that can serve as models. This research displays the following important characteristics:

(a) The goal was to prevent the development of reading difficulties in primary school. This was in contrast to the usual procedure which focused on the use of special techniques to treat existing reading difficulties. It has been shown that the latter procedure is expensive, complicated and often fails to produce the hoped-for result.

(b) A battery of reading-readiness, school-readiness and diagnostic reading tests for beginning reading were developed. Primary grade students were screened using reading and spelling tests given more frequently than usual. This was done since it had been demonstrated that the predictive value of these kinds of tests clearly decreases when they are administered at longer intervals.

(c) Children who, on the basis of the screening research could be expected to manifest reading difficulties, received special help from the reading teacher but at the same time continued to participate in the normal classroom or learning group.

The Dutch project (Van Dongen, 1984) followed these guidelines. In order to prevent reading difficulties to the greatest degree possible, a strategy was applied in three stages:

(a) The first step was to investigate to what extent it is possible to predict reading achievement before formal reading instruction starts. This is necessary in order to select the children who need special attention.

(b) In the second stage the teacher needed instruments and procedures for screening the pupils. The teacher had to assess reading readiness and reading achievement in initial reading.

(c) Further special care was given to the at-risk children to prevent the development of reading difficulties. Adequate educational procedures and programmes were necessary to stimulate the development of these children.

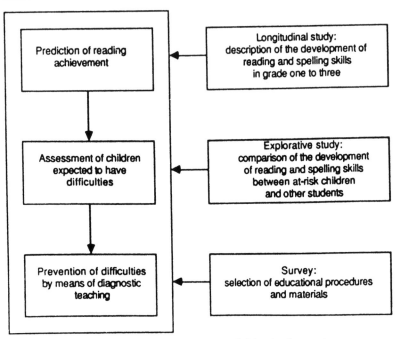

Figure 2: Stages and relations between the activities in the project.

Different activities were carried out, corresponding to the three stages.

1 A longitudinal study was done, in order to gather information about the development of reading and spelling skills in grades one to three.
2 An explorative study compared the reading development of the at-risk children with that of the other students.
3 A survey of educational procedures and materials available was made with the purpose of helping the teachers to select what would be most appropriate in a given situation.

We confine ourselves here to the longitudinal investigation into the development of reading comprehension and spelling skills in the first three grades of primary school.

1.5.4 LONGITUDINAL INVESTIGATION

1.5.4.1 General

In reading, a distinction is made between decoding skills and reading comprehension. Decoding skills refer to the ability to sound out written words and sentences. Reading comprehension requires a more thorough assimilation of the text on both the syntactic and the semantic level. In spelling, spoken language is decoded into graphic symbols according to a number of orthographic rules. This investigation had two aims:

1 to study the development of various sub-skills in word recognition, reading comprehension and spelling over a period of three years;
2 to predict reading and spelling scores from a comprehensive battery of tests administered at the beginning of formal reading instruction.

1.5.4.2 Research Questions

More specifically, we shall try to answer four questions.

1 The first question may be sub-divided into two separate questions:
    (a) Is the distinction between general and specific reading prerequisites relevant to the prediction of reading and spelling achievements?
    (b) To what extent does this distinction contribute to the prediction of the expected development of reading skills?

2 The second research question concerns the extent to which empirically distinguishable aspects are to be found in the development of reading and spelling skills after 3-4 months of formal reading instruction. The question is, in particular, whether it is possible to make a distinction between the ability to sound out words (power aspect) and the speed with which this process of decoding takes place (speed aspect), the power aspect being considered a precondition of the speed aspect.

3 The relations between decoding skills, reading comprehension and spelling after 8 and 12 months of formal reading instruction are the topic of the third research question.

4 The fourth research question concerns the longitudinal interrelations:
    - What are the direct and indirect relations between the two aspects: school readiness (general and specific reading prerequisites) at the outset of formal reading instruction, and the subsequent development of reading and spelling skills?
    - To what extent can decoding skills, reading comprehension and spelling be distinguished empirically as separate factors at the various points of measurement, and, especially, how strong is the influence of each one of them on itself in the course of time?
    - Do factors at one point of measurement influence factors of a different kind at a later point of measurement?

1.5.4.3 The sample

Two samples were investigated, each consisting of pupils from the first grade of 12 randomly selected schools. The first group used the basic reading programme 'Veilig Leren Lezen' (Learning to read safely); the second group used the programme 'Letterstad' (Lettertown). Because results from the two groups largely coincided, we are reporting only the results of the first one here (Mommers et al, 1986).

At the start of the first grade the total number of pupils belonging to this group was 310. At the beginning of the fourth grade, 225 pupils remained. Most of the drop-outs did so because they moved from the locality, but some pupils had to repeat a class or were consigned to special schools. The conclusions, therefore, only hold good for pupils not repeating a class.

1.5.4.4 The measuring instruments

In Table 1, a summary is presented of the measuring instruments used from the beginning of the first grade, up to and including the start of the second grade. The letters between brackets refer to factors on which the variables in the postulated starting model (Figure 3) bear. P is Prerequisites, DS is Decoding Speed, PO is Power aspect of reading and spelling, RC is Reading Comprehension and SP is Spelling.

In Table 2 a summary is presented of the measuring instruments used from the middle of the second grade up to and including the third grade.

Table 1: List of measuring instruments used in the first grade and at the beginning of the second. The letters between brackets refer to factors on which the variables in figure 3 bear.

measuring instrument (FACTOR)

A Point of time 1A
  Before the onset of formal reading instruction

1 Eli, subtest matrices	(P)
2 Eli, subtest copying forms	(P)
3 Rating school-readiness kindergarten teacher	(P)
4 Lettercluster identification test	(P)
5 Blending test	(P)
6 Phonemic segmentation test	(P)

B Point of time 1B
After 4 months of formal reading instruction

7 Caesar One Minute Test		(DSO)
8 Reading comprehension test		(PO)
9 Beginning reading test		(PO)
10 Spelling, -words VLL1		(PO)

C Point of time 1C
After 8 months of formal reading instruction

11 Caesar One Minute Test		(DS1)
12 AVI reading speed (text)		
13 Reading comprehension 1A CITO		(RC1)
14 Spelling, -words 2 CITO		(SP1)
16 Spelling, -sentences B1		(SP1)

D Point of time 2A
After 13 months of formal reading instruction second grade

17 One Minute Test A, Brus, Voeten		(DS2)
18 AVI reading speed (text)		(DS2)
19 Reading comprehension 1B CITO		(RC2)
20 Spelling, -words OBCE 1		(SP2)
21 Spelling, -sentences B2		(SP2)
22 Spelling, -sentences C1		(SP2)

Table 2: List of measuring instruments used from the middle of the second grade up to the end of the third grade. The letters between brackets refer to factors on which the variables in the postulated model bear.

measuring instrument   (FACTOR)

E Point of time 2B
After some one and half year of formal reading instruction

23 One Minute Test A; Brus, Voeten		(DS3)
24 AVI reading speed (text)		(DS3)
25 Reading comprehension 2 CITO		(RC3)
26 Written assignments 2		(RC3)
27 Spelling, -words OBCE2		(SP3)
28 Spelling, -sentences C2		(SP3)
29 Spelling, -sentences D1		(SP3)

F Point of time 3A
  After some two years of formal reading instruction (beginning of
  the third grade)

30	One Minute Test A; Voeten, Brus	(DS4)
31	AVI reading speed (text)	(DS4)
32	Written assignments 3	(RC4)
33	Spelling, -words OBCE2	(SP4)
34	Spellin, -sentences D2	(SP4)
35	Spelling, -sentences E1	(SP4)

G Point of time 3B
  After some two and a half years of formal reading instruction

36	One Minute Test A; Brus, Voeten	(DS5)
37	AVI reading speed (text)	(DS5)
38	Written assignments 4	(RC5)
39	Reading comprehension M3 CITO	(RC5)
40	Spelling, -words OBCE3	(SP5)
41	Spelling, -sentences E2	(SP5)
42	Spelling, -sentences F1	

1.5.4.5 The analyses

All the analyses in this study were executed by means of the Lisrel programme, version VI (Joreskog and Sorbom, 1981). The reason for the choice of the Lisrel method is the opportunity this approach offers to evaluate by means of the maximum likelihood method, both the global and local fit of models in which both cross-sectional and longitudinal effects may be included, and the possibility of employing observed, as well as latent variables. The correlation-matrices of the measured variables were taken as input to run the programme on.

A distinction should be made between the period of beginning reading (up to the start of the second grade) and the period of transition from beginning reading to developmental reading. Research questions one to three refer to the first period, and the fourth question to the second period. This distinction was why the data were analysed in two stages. As a first step those models were analysed that relate to the first period only. At a later stage, the data up to and including the third grade were added.

1.5.4.6 The postulated starting model

When designing a model for longitudinal data, the time factor is obviously of great importance. It is clearly quite impossible to assume that variables later in time could have any influence on those earlier in time. Apart from that, the following theoretical considerations, which constitute a further elaboration of the ideas mentioned in section 2, led to the design of the starting model represented in Figure 3:

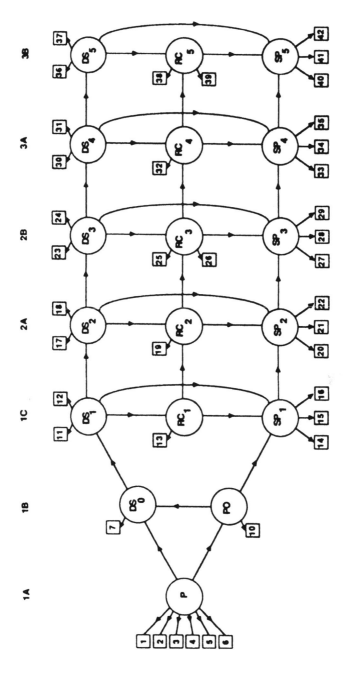

Figure 3: Starting model. The numbers in the squares refer to the tests presented in Tables 1 and 2.

(a) The reading prerequisites measured at the start of the first grade (point of time 1A) either directly or indirectly influence all variables later in time.

(b) After three to four months of formal instruction (point of time 1B) the reading tests are subdivided into speed and power tests. Since at this stage the reading speed depends on the precision with which words are sounded out, an influence of the power aspect (PO) on the speed aspect (DSO) is postulated.

(c) The speed aspect continues in Decoding Skills (DS), the power aspect is broken up into Spelling (SP) and Reading Comprehension (RC). In the course of the first grade the process as described under (b) is actually reversed: Reading Comprehension and Spelling are considered to be determined by the decoding speed, as measured in Decoding Skills. This is seen in relation to the limited capacity of the working memory, when coding and decoding processes proceed more or less automatically, on account of which more attention may be paid to other processes. It also seems plausible to assume that at such an early stage the comprehension of words and word-structures has a positive effect on spelling achievements.

(d) Apart from the cross-sectional effects, the longitudinal influences concerning the three separate factors of Decoding Skills, Reading Comprehension and Spelling play a prominent part in the model.

In Figure 1 both the measured variables (squares) and the latent variables (circles) have been marked. The numbers refer to Tables 1 and 2.

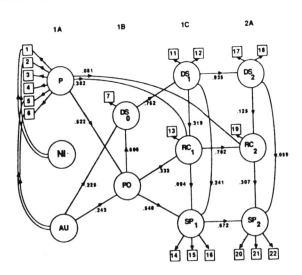

Figure 4: LISREL model based on one general factor, Prerequisites, and two secondary factors, an Auditory factor and a Non-verbal Intelligence factor.

1.5.4.7 Results and conclusions

In the starting model (see Figure 3) one general factor was postulated, which was labelled Prerequisites (P). It was encouraging to see that most of the coefficients were positive and rather high. However, the modification indices in the output indicated several problems. Two of them appeared in the theta-delta matrix. The first one concerned the non-verbal intelligence tests (subtests matrices and subtest copying forms), the common factor, labelled Prerequisites, and two special factors, namely an auditory factor and a rest factor related to the intelligence tests (see Figure 4).

This seemed to be a considerable improvement. In addition, the modification indices indicated a direct influence of the Auditory factor (AU) on the factor Decoding speed (DS) and on the Power factor reading and spelling (PO). Furthermore, it was clear that the direct influence of the common factor Prerequisites (P) on Decoding Speed, which had been postulated in the starting model, appeared only indirectly through the Power factor (PO).

It would carry us too far to describe all the attempts that were made to improve the model. Figure 4 shows the model which fits best. When Figure 4 is compared with Figure 3, it is remarkable that the common factor Prerequisites (P) directly affected the factor Reading Comprehension (RC1) with a rather high path coefficient (.382).

In accordance with the verbal coding efficiency model of reading skill as developed by Perfetti and Lesgold (1979), it was found that the factor Decoding Speed (DS) affected Reading Comprehension (RC) at the third and fourth points of measurement. The coefficients were, however, although significant, rather low.

The direct effect, postulated in the starting model, of Decoding Speed (DS) on Spelling (SP) was limited, as was the effect of the factor Reading Comprehension (RC) on Spelling (SP).

This set of results responded to the first three research questions. Figure 5 gives an overall answer to the fourth question.

The influence of the Prerequisites (P) on Reading Comprehension decreases in the third grade (3A en 3B). From the high path coefficients for the effects of the factors on themselves, it may be concluded that Decoding Skills, Reading Comprehension and Spelling, starting from the point of measurement 1C, are clearly distinguishable factors. The distinctive character of these three factors is revealed much more clearly than in cross-sectional correlation research.

After point of measurement 2A the interconnections between Decoding Speed, Reading Comprehension and Spelling do not show a very consistent picture. This was also the case with the Lettertown-group.

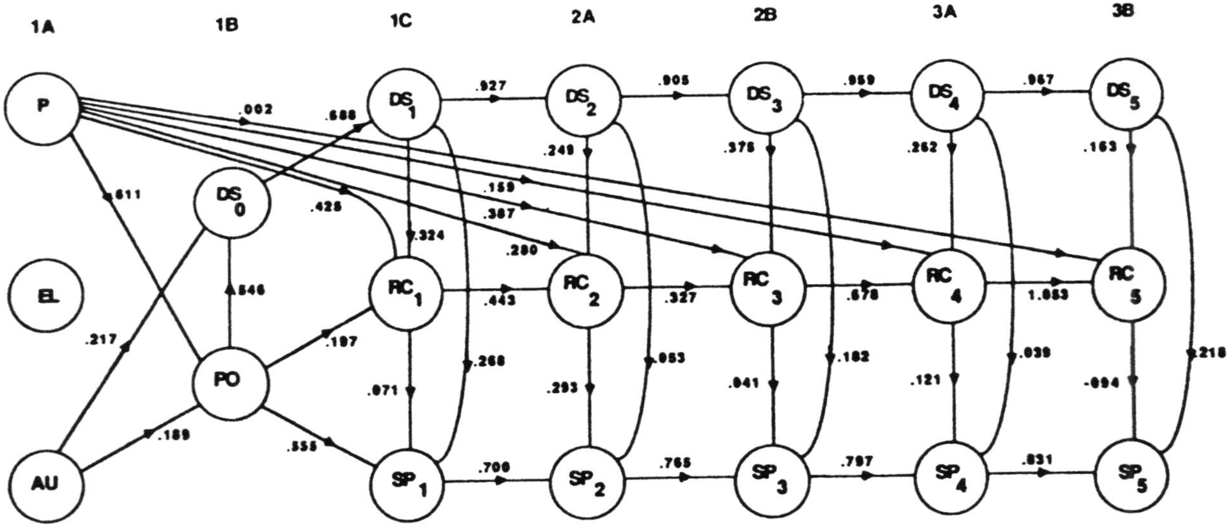

Figure 5: An overall picture of the path coefficients between the factors in the first three grades of primary school.

These results are of importance for the planning of course material, the formulation of objectives and goals and the diagnosis of reading and spelling difficulties. In an integrated course of formal reading instruction, a fair amount of attention will have to be paid to the specific nature of decoding skills, reading comprehension and spelling. Pupils with reading and spelling difficulties will have to be examined on all three of these factors, because their relative independence means that discrepancies may exist between the stages of development of the respective skills. Although the variables were measured fairly globally, the clear relations that have nevertheless been found in this longitudinal research are remarkable, both from a theoretical point of view and from the practical point of view of formal reading instruction.

It will be clear that these research results belonging to the lower plane in Gideonse's model are also relevant to the two upper planes. They in no way pre-empt further discussion on this topic, but we now know a little more about the picture than we did before this longitudinal study was carried out.

REFERENCES

Dongen D van (1984) *Leesmoeilijkheden. Naar diagnostiserend onderwijzen bij het leren lezen.* Tilburg: Zwijsen.

Gideonse H D (1968) An output-oriented model of research and development and its relationship to educational improvement. *The Journal of Experimental Education,* 37, 157-163.

Joreskog K G & Sorbom D (1981) *Lisrel. Analysis of Linear Structural Relationships by Maximum Likelihood and Least Squares Methods.* Uppsala: University of Uppsala.

Malmquist E (1969) *Lassvarigheter pa grundskolans lagstadium.* Experimentalla studier. Falkoping.

Malmquist E & Valton R (1974) *Forderung legasthenischer Kinder in der Schule.* Weinheim: Beltz Verlag.

Mommers M J C, Leeuwe F J F van, Oud J H L & Janssens J M A M (1986) Decoding Skills, reading comprehension and spelling: a longitudinal investigation. *Tijdschrift voor Onderwijsresearch, 11 (2),* 97-114.

Perfetti C A & Lesgold A M (1979) Coding and comprehension in skilled reading and implications for reading instruction. In L B Resnick & P Weaver (Eds). *Theory and Practice of early reading (Vol. 1)* Hillsdale, N J: Erlbaum.

ACKNOWLEDGEMENTS

This study was sponsored by the Foundation for Educational Research (Stichting voor Onderzoek van het Onderwijs). The author would like to express his gratitude to J van Leeuwe, H Out, J Janssens, D van Dongen and H Hulsmans for their continued support and efforts during the conduct of this study.

1.6 DIFFERENTIATED EARLY READING INSTRUCTION: EXPERIENCES WITH A FLEMISH TEACHING-LEARNING PACKAGE

by

A JANSSENS and E DE CORTE, Belgium

1.6.1 SUMMARY

Starting from dissatisfaction with the typical situation of early reading instruction in Flemish schools a few years ago, a group of inspectors and teachers began constructing a teaching-learning package for language instruction in the primary school. The major characteristics of the teaching of reading in the first grade, based on this package are: systematic differentiation of instruction, emphasis on children's experiences as a starting point for learning and teaching, and explicit attention to enhancing the motivation to read. The present paper identifies some major problems of present-day typical reading instruction, and describes briefly differentiated reading instruction according to the new language package. Afterwards, some findings in children as well as in teachers are reported that derive from implementation of this method of teaching reading in a substantial number of first grades. The children seem to achieve better reading results, and they show a high motivation for reading in, as well as out, of school. For many teachers the transition from whole-class teaching toward differentiated reading instruction is quite difficult. The paper also discusses two ways in which the teaching-learning package tries to cope with teachers' problems, namely the organisation of in-service training sessions, and the availability of a well-elaborated teacher manual.

1.6.2 SOME ASPECTS OF TYPICAL READING INSTRUCTION

In most first grades of the primary schools in Flanders, reading instruction still occurs in the form of whole-class teaching. This means that all pupils of one class are simultaneously taught the same content, in the same way, by one teacher. Observation over a longer period of time of such a class of young readers reveals a rather striking change in children's activity. In the beginning they are all strongly motivated; however, after some time the interest and enthusiasm of a number of pupils decrease rather quickly.

Another fact relates to the situation of the poor readers. Although most first grade teachers see possibilities for helping poor readers within their classroom, only a few of them are successful in implementing possible measures. The typical situation is to practise the content once again in the same way as during whole-class instruction. The results of a recent study on poor learners in the first grade are in line with this characterisation of instructional practice. As an illustration, we summarise the major findings of this investigation in this respect (1984).

- 86% of the 102 teachers who participated in the study mention possible measures for helping the poor readers in their class.

- 67% of the participants state that they help children through measures with respect to the nature of the content of instruction. However, only 50% of the teachers specify this statement; mostly this relates to simplification or reduction of the content.

- 47% of the teachers say that they help children through modifications in their method of instruction, but only 33% of the participants are more specific about this aspect.

- Finally, 29% of the teachers mention adaptive evaluation as a means for helping poor readers; only 18% specify what this involves (in seven cases it seems to be a form of individual evaluation).

Starting from this finding and from everyday observation in classrooms, it can be stated that a thorough diagnosis of the nature and the possible causes of children's errors and difficulties is very frequently lacking in today's classrooms. Often teachers seem not to be aware of the importance of such a diagnosis as a starting point for remediation, and, moreover, they seem to lack the necessary diagnostic skills.

Learning to read is a very complex process (see for example Calfee and Drum, 1986), requiring much effort from the child as well as much skill and understanding from the teacher. Therefore, it is not surprising that teachers tend to focus very strongly during whole-class instruction on the micro-aspects of the reading acquisition process, thereby neglecting other important factors such as differences in children's language proficiency, in their expectations with respect to reading instruction and in their fear of failure. These factors can directly influence children's motivation, and consequently have an impact on the process of learning to read.

1.6.3 CONSTRUCTION OF A NEW TEACHING-LEARNING PACKAGE FOR READING INSTRUCTION

1.6.3.1 General

Taking into account the preceding data concerning the typical situation of reading instruction, a group of inspectors and teachers started, a few years ago, constructing a teaching-learning package for language instruction in the primary school. It goes without saying that in the first grade, acquisition of the reading technique is a central issue in this package; however, in addition, considerably more attention than in the typical Flemish textbooks for reading instruction is paid to the following aspects: differentiation of instruction, emphasis on children's experiences as a starting point for learning and teaching, and explicit attention to enhancing the motivation to read (Rotthier, Van Hul and Van Puyvelde, 1984). Thereby the developers of the package had the following objectives in mind:

- enriching children's language proficiency in view of an appropriate approach of the reading materials to be studied;

- differentiated presentation of the reading materials, using a variety of texts, in order to stimulate the motivation to read. In this respect some activities are inserted to show the children the functional value of reading;

- evaluation of the learning objectives, taking explicitly into account the built-in differentiation;

- promoting a change in the teaching behaviour of the users, especially enhancing the skill in diagnosis and remediation of difficulties in poor readers. During the try-out stage of the package it seemed that, in view of this objective, the teachers needed frequent support and assistance.

In the rest of the present contribution, we will not address the micro-processes of learning to read, but will limit ourselves to a discussion of the other influential factors involved in reading acquisition that were mentioned above.

1.6.3.2 Enriching children's language proficiency

The package is elaborated around topics that are derived from children's everyday life and experience. It is well-known that pupils' antecedent experiences with respect to the content of reading materials vary strongly from very rich to extremely poor. In order to ensure that all children can approach the reading materials with equal opportunities, each topic starts with a common "entry experience". In that context the teacher's manual suggests a number of alternatives, from which the teacher can choose in consultation with his pupils (e.g. a short walk in the forest, an observation period in the surroundings of the school, or possibly in the school). During this "entry experience" the teacher tries to activate as many sensory organs as possible, and she listens to the spontaneous questions and remarks of the pupils before asking questions herself. It is also suggested that useful things and evidence are brought into the classroom, e.g. objects, sounds, pictures, slides, interviews.

The "entry experience" is followed by a classroom discussion focused on a series of open questions that provoke children's interest, stimulate their motivation and promote independent thinking. During this discussion the real world is taken as the starting point: concrete things and their "transformation in language" remain connected during a sufficiently long period of time. Consequently, the "entry experiences" constitute a solid basis for the information and contents presented next in speech and text. There is a continuous interaction between the child's own language utterances, and the contents presented in speech and text.

1.6.3.3 Differentiated reading instructions

Every week, two lesson periods of half an hour are scheduled for differentiated reading. With respect to each topic, three reading texts are presented: a basic text which is used during whole-class instruction, a simpler one for the poor readers, and an enriched text for the skilled readers (with separate booklets for each text). Additional reading exercises are availble for those pupils who have perfect command of the enriched texts; they have to carry out these additional exercises independently.

The decision to provide three texts with respect to each topic was taken in order to represent different levels of difficulty, but also with a view to stimulating children's motivation to read. Indeed, it is a remarkable phenomenon that almost all pupils enter the process of learning to read with enthusiasm, but that a lot already show some "reading fatigue" at the end of the first grade, if not before. And it should be added that this is not only observed in poor readers. It is our opinion that instruction is probably at least partly responsible for this decline in motivation. Indeed, to practise a reading text it is very common for the teacher to give individual children alternately a turn to read aloud a part of the text, while the whole class has to follow. Following this procedure, the same text is read up to four or five times. An attentive observer will notice signs of boredom even when the text is read the second time. It is not surprising that children who have experienced this kind of reading practice for months, and even years, progressively lose their initial enthusiasm about reading. Although most children at the end of the sixth grade master the technical skills of reading rather well, it seems to us that the primary school fails to achieve the major goal of reading instruction if twelve-year-olds to not spontaneously use and enjoy books for acquiriing new knowledge, as well as for recreation.

The teacher's manual contains a lot of suggestions concerning diagnosis and remediation of the difficulties of poor readers. Every two weeks reading tests are administered in order to determine children's reading level; consequently the groups representing the three levels are very flexible.

The practical organisation of the differentiated reading periods differs according to the specific situation and resources of each school.

- Some teachers manage to run the activities all by themselves. During a differentiated lesson period they work with three groups representing the three reading levels. Moreover they combine homogeneous and heterogeneous grouping. Homogeneous groups always work under the direct supervision of the teacher. For example, while the teacher is busy with the poor group, the pupils of the other two groups form pairs: a skilled reader tutors an average reader.

- In schools where there is a remedial teacher, the latter teaches the poorest readers during a differentiated reading period, while the class teacher works as mentioned above.

- When there is no remedial teacher available, the principal of the school or a so-called "ambulant" teacher assists the class teacher.

- In schools with a well-functioning parent association, some parents operate as teacher aids during differentiated reading lessons.

- Finally, in some schools the class teacher, the remedial teacher and one or more parents co-operate simultaneously in one classroom.

1.6.4 SOME FINDINGS WITH RESPECT TO DIFFERENTIATED REA-DING INSTRUCTION

1.6.4.1 General

In this section we report some findings derived from implementations of differentiated reading instruction according to the new language package. These findings are based on more than 300 observations of lessons in classrooms, and on discussion sessions with groups of teachers (about 1000 individuals), all spread over the past five years.

1.6.4.2 Results observed in children

First of all, it seems that children continue to like reading, even out of school. This is probably due to having experienced success in reading texts that are adapted to their level of proficiency, as well as to variation in the reading materials. In any case, children's motivation to read seems to be high.

The skilled readers actually help the others, which allows the teacher to spend more time with the poor readers. The reading materials are presented to the latter group in very varied (game-like) formats; as a consequence, the content also appears as being new to children of the other groups. During the evaluation sessions with the whole class, the poor learners also have the opportunity to show that they are successful readers; this seems to have a very stimulating effect. The frequent occurrence of reading in dyads increases considerably the reading opportunities of the poor readers.

It is suprising how much progress these young children make independently, they seem able to concentrate via a series of individually adapted reading tasks, while the teacher is working simultaneously in the same room with another group. Moreover, there are indications that this readiness for independent task-oriented behaviour generalises towards other domains.

Finally, test results show that the individual differences in reading ability are accentuated as a consequence of instruction based on the package. The skilled pupils out-perform the good performers in classes in which the package is not used; however, the reading performance of the poor readers also increases, although not so much. An increase in individual differences is indeed what one would expect as a result of appropriate and effective individualised instruction. Obviously this has implications for the task of the teacher, because it requires skill in organising classroom activities as well as the availability of a broad selection of learning materials.

1.6.4.3 Problems observed in teachers

First, it seems that for many teachers the transition from whole-class teaching towards differentiated reading instruction is quite difficult. Although they are convinced of the large differences in reading ability in their class, they have a lot of problems in modifying their teaching to take account of these differences. Probably arising from feelings of insecurity, a lot of teachers initially advance counter-arguments against the new ideas, such as: "first graders are unable to work independently"; "they will be disturbed while I work with another group", etc. The developers of the new package were aware that good differentiated reading instruction requires a substantial modification in the usual teaching behaviour. Therefore, they were not surprised to see, initially, a regression towards traditional whole-class instruction. Indeed, some teachers work with different groups, without changing their teaching style; they gave, as it were, a whole-class lesson three times.

Second, working with smaller groups confronts the teachers more quickly and more acutely with the problems of individual learners. This leads to the awareness of shortcomings in one's own teaching skills, and knowledge about the process of reading and learning to read; consequently, the feelings of insecurity mentioned above may arise, possibly followed by a regression to the old habits. In those who persist, one observes a growing need for information and assistance.

1.6.5 HELPING TEACHERS TO SOLVE THEIR PROBLEMS WITH DIFFE- RENTIATED READING INSTRUCTION

1.6.5.1 General

The developers of the teaching-learning package for language instruction try to help teachers with their problems related to differentiated reading instruction in two different ways, namely the organisation of in-service training sessions, and the availability of a well-developed teacher manual.

1.6.5.2 In-service training

For a sample of 25 schools which started with differentiated reading instruction, a series of ten in-service training sessions was organised, spread over a two-year period. Participants in these half-day sessions were the class teachers of the first grades, the remedial teacher, and the principal of each school. The total group of 75 persons was subdivided into three groups of 25.

At the first stage, the teacher's need for more information concerning the reading process was met. During three sessions, the different components of the processes of learning to read and to write were thoroughly studied and discussed. In the first two sessions, data about reading lessons attended were analysed in combination with a more theoretical approach. The third session was mainly devoted to an analysis of classroom experiences, with differentiated instruction in terms of the different components of the processes of learning to read and to write. It seemed to be a good decision to involve the first grade class teachers, the remedial teacher and the principal together in the training. This had a positive impact on co-operation and mutual support in their school. Together they also took care of the transmission of information to, and for the continuation of, the approach in the second grade.

At the second stage (two sessions) several classes were visited in which differentiated reading was applied skilfully and fluently. This allowed the participants to observe for themselves that first graders can work independently on a task very well, that reading in dyads progresses smoothly, and that, by using a well-elaborated evaluation schedule, the class teacher can follow up children's progress very precisely. Through discussions with the teachers of the classes that were visited, the participants recognised their own situation and problems; but the evidence provided by colleagues seemed to be very convincing to continue with differentiated reading instruction.

The five sessions of the third stage were devoted to discussions of a series of data observed and collected by an external consultant during his visits to the schools and classes of the participants in the in-service training. The discussions focused on the following topics.

- Critical analysis of the structure of reading lessons, taking into account the intended objectives.

- Critical analysis of the types of exercises used. Frequently, teachers discovered here that certain forms of exercises, which they had applied for years, were not in accordance with the objectives. For example, certain tasks were actually spelling exercises instead of reading exercises.

- Analysis and categorisation of reading errors, and definition of a strategy for remediation. This activity led to an inventory of frequently occurring errors, complemented by a bank of exercises.

- Comparison of the new approach to reading instruction with the method applied before. This led to a clear identification of the shortcomings of the latter method. A good example in this respect is that, in general, not enough attention is paid to auditive analysis and synthesis during early reading instruction.

Generally speaking, the results of this in-service training were very positive. Nevertheless, the external consultants came to the disappointing finding that, in spite of the training, quite a number of teachers were still unsuccessful in correctly diagnosing frequently-occurring reading errors, and in proposing an appropriate strategy for remediation. One wonders what happened to the poor readers in their classes during all the years before!

1.6.5.3 A well-elaborated teacher manual

In our opinion, a good manual accompanying a teaching-learning package should be a book of reference for the teacher, without, however, curtailing her creativity. It should inform her about the intentions of the package, and contain possible answers and solutions to her questions and problems. Moreover, the manual should give sufficient background information that underlies and justifies the instructional approach.

The manual accompanying the present language package has been substantially improved and enriched by the data and experiences collected during the try-out stage and during the in-service training sessions. The following aspects are discussed extensively: frequently-occurring problems of teachers; the different components of the process of learning to read; learning objectives, teaching strategies, possible difficulties in children, and suggestions for their remediation.

The intended forms of differentiation are described in terms of lesson schedules specifying the possible contribution of the remedial teacher, the ambulant teacher, and other teacher aids such as parents. As already mentioned, level-adapted reading tests are scheduled for every two weeks, and the manual provides appropriate suggestions for possible remediation.

On the basis of the observation of, and the experience with, the implementation of the language package over the past years, the developers claim that a well-elaborated manual containing specific and detailed support for users, accompanied by sufficient background information, counterbalances strongly the regression of teachers towards purely whole-class instruction.

1.6.6 CONCLUSION

It is evident that continued research aiming at a better understanding of the micro-processes involved in learning to read, remains a very important task. However, on the basis of observations and experiences within a developmental context, we have argued that these micro-processes are strongly influenced by a number of other factors situated comparatively at a macro-level, such as: children's background, motivation and attitudes; the structure of the class group; the teaching methods used and the forms of differentiation applied; the instructional skills, the experience, and the attitude of the teacher. It is our conviction that further investigation of the global context of, and the activities during, reading instruction can be very illuminating with respect to the reading difficulties of a substantial number of children. In this respect we would like to make a plea for carrying out a series of teaching experiments: using the underlying ideas and principles of the language package discussed in this paper, experimental teaching-learning units could be designed and implemented in studies focusing on specific instructional and situational variables, and aiming at a more through understanding of their effects on the process of learning to read. In the meantime, we dare already state that the practice of reading instruction can take advantage of the approach underlying this language package. Improvement of the teaching of reading in line with this approach seems to produce the following results: teachers get a better understanding of the reading process, and they are given more support in view of effective diagnosis and remediation of children's difficulties. The children themselves seem to achieve better reading results, and, what is evey more important, their reading motivation is maintained — often even increased.

REFERENCES

Calfee R and Drum P (1986) Research on teaching reading. In M C Wittrock (Ed), *Handbook of research on teaching* (Third edition) (pp.804-849). New York: Macmillan.

CSPO (Centrum voor Sociaal en Psychopedagogisch Onderzoek) (1984). *Zwakfunctioneren in het eerste leerjaar van de lagere school. Rapport 6: Klasmilieu-onderzoek eerste leerjaar.* Leuven: CSPO.

Rotther G, Van Hul R, and Van Puyvelde B (1984) *Taal voor vandaag en morgen: Leen en Rik.* Leuven: J B Wolters.

1.7 INVESTIGATIONS IN LEARNING TO READ AND TO WRITE IN CLASS 1 AS FOUNDATIONS OF TEACHING METHODS FOR ELEMENTARY SCHOOLING

by

Prof. Dr. Mechtild DEHN, W.Germany

1.7.1 SUMMARY

Insight is given into a long-term study regarding the acquisition of reading and writing skills. Research into primary reading has up to now faced this actual process: our knowledge has been orientated on the one hand towards the analysis of the structure of written language, and on the other hand towards examinations of the reading and writing of the advanced learner or towards the learning difficulties of older pupils. The criteria which are set out here in order for discussion on an interpretation of learning processes at school, are well-founded in the psychology of thinking and in the theory of cognition. They can help to identify difficulties in learning at an early stage. The results cast some doubt on current ideas about the relationship between teaching and learning.

1.7.2 INTRODUCTION

In the first decades after 1945, research on teaching methods for beginners was characterised by the discussions regarding the effectiveness of teaching procedures[1] and regarding the contents, the language and the kind of exercises in primers[2]. However, only in the last ten years has interest focused on the procedure of the child when acquiring the ability of writing and reading. That our aims of research are now centering on the process of the learner's perception and the forming of rules in his mind has been mainly caused by two developments since the mid-seventies in Germany: one of these has been started due to the strong

---

[1]   Schmalohr (1961), Ferdinand (1970), Weinert *et al* (1966).
[2]   Bauer (1971), Geiss (1972), Gromminger (1970), Merkelbach (1974), Hannig/Hannig (1974), Dehn (1975), Menzel (1975), Rigol (1973), Bergk (1980), Dehn (1977).

criticism of the term dyslexia and the methods of its examination[1], where the other one is due to studies outside of the Federal Republic of Germany, mostly in the field of the American cognitive psychology INeisser 1979) and psycholinguistics (Goodman 1976), and the structural and generative theory of language as well as the Russian psychology of learning[2].

The *research results* that I want to present here belong to this theoretical frame. They try to describe and to understand the *acquisition* of reading/writing skills from the angle of the child. To comprehend *learning how to read* as a cognitive development proved to be informative as well as studying it as a *process of solving problems.* By examining *learning to write* it has been discovered that here the child accomplishes *an important act as to the analysis of the language,* for instance when he learns to write down what he has formulated in his mind whilst simultaneously he gradually picks up the rules of orthography[3].

In view of the methods of research, it turned out that for many questions it is useful *to analyse the mistakes,* which is an instrument provided by the learner himself: the mistakes the child makes when he writes — particularly when he writes "spontaneously" — as well as when he reads aloud (necessary for him as a beginner) allow an insight into cognitive processes. Till now the mistakes have been regarded primarily as deviation from the taught norm, as missing the correct result (cf. in this connection the typology of mistakes by R Muller, 1974), whereas, at present, mistakes — chiefly in view of the primary acquisition of language — are regarded as a "necessity pertaining to the development" (Wode 1978). I propose to regard *mistakes as a necessity of learning.* The analysis of mistakes is typical not merely in view of the latest studies on the acquisition of writing/reading skills but is also used in applied linguistics (German as a second language, linguistics of foreign languages) and didactics of mathematics. The analysis of mistakes offers the chance — independent of experimental arrangements (tachistoscope: filming of the eye movement) that can hardly be expected of reading beginners — to observe the learning situation itself, i.e. we are no longer left to conclusions regarding the *learning* process by means of examining the process of reading and writing which, of course, is limited as to its cognitive value. The instrument of mistake analysis should not be over-rated, however, since not all mistakes can be interpreted in this way — some emerge quite accidentally.

For the research team under my direction, the *aim of analysing the acquisition* of reading and writing skills in view of the development of cognitive processes is to discern the difficulties in learning with more details and earlier than up to now, and to apprehend these in view of their position in the process of learning. Another aim of our research is to provide a basis for pedagogical and didactical decision regarding the curriculum and in view of individual help.

---

1   Schlee (1976), Sirch (1975), Spitta (1977).
2   Hofer (1976), Weigl (1974), A A Leont'ev (1975).
3   Brugelmann (1984), Balhorn (1983, 1985), Dehn (1984, 1985), Downing/Valtin (eds) (1984), May (1986), Huttis (1981, 1985), Weinert/Kluge (eds) (1984).

1.7.3 REPORT UPON A LONG-TERM STUDY

1.7.3.1 General

One important problem of analysing the acquisition of writing/reading skills at school refers to the fact that the individual process of learning is influenced on the one hand by the starting conditions of learning (resulting from processes of learning during the time before schooling begins) and on the other hand by the teaching procedure/method affecting all pupils of a class. Due to the fact that the instruction offered is digested in different ways among other things due to the 'a.m.' conditions — the individual beginner in reading is learning less or more, and in any case other things too, than that which the teacher teaches — we have a highly complicated system of teaching procedures and learning processes. The main point is not the interpretation of the learning processes — imagined as almost indigenous — but the different adaptation of the teaching procedures.

This means, for the formulation of tasks for long-term studies, that a task cannot be simply repeated (e.g. to find out the degree of approach to the correct result in the course of time), but that the task has to be related to the position in the course of instruction and this relation has to be taken into account when evaluating. Another methodological problem refers to constructing the examination situation. If individual children were to be observed during lessons, most of the cognitive processes cannot be grasped. In order to focus on the process of learning, and to maintain the fundamental relation between teacher and learner in the situation of observation, we decided in favour of observing the individual child: in the course of the first scholastic year, a pedagogically acting test administrator meets the beginner in reading and writing several times, gives the tasks and observes as well as helping to accomplish these. This situation is quite natural at school and not at all artificial. It resembles the teacher-learning situation in examples of interior differentiation according to achievement in the classes. The beginner in reading also meets with this situation when he has to solve a reading problem by himself and someone is helping him in case of need. The job of the tester is to encourage and to support the child by repeating what he has said already, by answering his questions and by giving general advice, like "Start once again from the beginning!". The reading results of the children, as well as the comments of the tester, have been noted and evaluated.

In 1979/80 and 1981/82, 66 pupils of seven classes at Hamburg schools were observed in the course of their first scholastic year — in total every pupil ten times. The schools belong to different areas as far as social classes are concerned. The women teachers made use of two different courses of instruction. One of these is structured in a strictly linear way, stressing the blending from the very beginning, whereas the other allows elements of reading according to the "look-and-say" method at the beginning.

Each time of observation, the pupils were given a series of tasks: to read some words of the course, to read some unknown words, to read some texts, to blend some sounds, to segmentize each word from a chain of words written without intermediate space, to write some letters and even some words which are not dealt with in lessons, and to complete the last sentence of a text that had been presented orally. In this way we got a lot of observation data of every child: of the first grade, approximately 14,000 reactions in total, to singly presented words referring to a text, and 2,800 written words that had not been dealt with in lessons. Furthermore, we collected data on these 66 pupils regarding several standardised tests (Bremer Artikulationstest [BAT], Bremer Lautdiskriminationstest [BLDT], Heidelberger Sprachentwicklungstest [CFT], Schulleistungstestbatterie 1 [SBLI]), and CIT. We followed these 66 pupils to the end of their elementary schooling with a reading and spelling text.

1.7.3.2 Learning to read as the solution of problems

When evaluating (cf Dehn 1984) we found three levels, the relationship of which might alter the pedagogical comprehension of the acquisition of reading skills. The first two levels are known and often used in everyday life at school as well as in diagnostic instruments: this is, first, the mastering of the elements of writing, i.e. the graphemes, which means the material level; the second refers to the mastering of partial operations, particularly the skill of blending, which means the level of operating with these elements (cf. the term "tactics" of Brugelmann/Fischer 1977). The third level refers to the relation and the sequence of the single operations in the process of reading. In our view this level of "meta-operations" seems to be central to understanding the process of acquiring reading skills and the difficulties involved therein.

1.7.3.3 Three levels of analysing

(a) The material level of reading

It is quite remarkable that dealing with the graphemes in lessons does not play the role accorded to it. We have examined the knowledge of graphemes in three different ways:

We have

- analysed the receptive mastering of graphemes in the case of reading tasks (individual words, reading a text); these tasks always also included graphemes which have not yet been dealt with in lessons, at least in some of the classes;

- observed the productive handling when words had to be written that were virtually unknown for the most part ("write down 'Lampe'" [lamp]); also, these tasks included graphemes that had not yet been the subject of learning, at least for some of the classes;

- dictated three graphemes to the children during every observation, i.e. we examined to what extent the children could reproduce graphemes that had already been dealt with in all classes ("write down the small/capital letter...!").

90

In view of these different ways of mastering the graphemes we compared the position in the process of learning and the position in the course of instruction. Furthermore, we related the mastering of graphemes to the kind of approach of the beginners in reading.

In the case of the word "Hammer"[hammer]) to be read, the letter "e" had not been the subject of learning for 16 out of 32 children, but none inquired as to that grapheme. No reading process failed due to the missing/unknown "e" (receptive mastering). The same refers to "leise" [quiet]: besides the "e", the letter "s" had not yet been introduced to 16 children in their lessons. For the group of 81/82 the "m" of "Motor" had not yet been a learning subject (although the frequently appearing name of the figure MARIO in the primer starts with the same letter). Two of the 34 pupils questioned it; the rest did not have difficulties with this grapheme in the process of reading. (This refers to its first appearance in the text and to the immediately preceding single word as well — three children enquired as to the grapheme in this case).

In the case of the word "hat" [has], the three graphemes of which had been already introduced, 11 out of 32 children (i.e. 30%) confused it with the visually similar words "halt" and "holt" [catches] of the course — syntactically both could be possible, too. And at the beginning "Motor" often was mixed up with the word "Mario" of the teaching course — and that indeed is semantically and syntactically reasonable here, too.

This "independence" of the learners from the teaching actions on this material level becomes even more obvious when it comes to writing down unknown words (*productive mastering*): although "w" had not been introduced to the 34 children of the scholastic year 81/82, 18 of them wrote it down correctly in the word "Lastwagen" [truck]; the equally "unknown" "m" of "Arme" [arms] was written down correctly by 24 children, and the "p" of "Lampe" [lamp] was noted down by 16, although it had not been dealt with in the lesson.

The knowledge of letters (*reproductive mastering*) examined by means of dictation of letters does not develop continuously like the other skills which we observed, but in its means, too, it is subject to variations in the course of the scholastic year (cf. 1984, p.102). The same could be demonstrated in view of reading tasks and tasks referring to the writing of words. On the one hand the children already knew — probably due to extra-curricular learning processes — a number of graphemes which had not been dealt with at all, and on the other hand they did not master the subjects of the teaching course to the expected or desired extent. In view of the poor readers, the knowledge of letters is not constantly below the average; but even good readers are not certain regarding some of the graphemes which have already been dealt with.

Regarding the connection between the mastering of graphemes and the mastering of partial operations for reading a single word or a text, we found that the status/value of knowing the graphemes was different for good beginners in reading and for poor readers. Whereas good readers are able to compensate for their insufficient knowledge of letters (e.g. they enquire more often); for poor readers, the mix-up of graphemes often becomes an obstacle which they cannot get over. They become involved in lengthy attempts and new confusions. From our observations we think that the stress that many teachers lay on the mere teaching of these material items seems to be rather dubious when used to judge how far their classes or individual pupils had proceeded. However, the general mastering of a concept "letter" is indeed an essential criterion when acquiring writing skills. In the groups we examined, the small number of children who couldn't write or name a single letter when schooling started belong to those who had difficulties in learning to read and to write in their first four grades.

(b) *The level of partial operations/working on sections when learning to read.*

Teachers are assigning great importance to the ability to blend when judging the achievement of pupils. This is quite justified, and due to our investigations this ability is likewise a criterion for lasting difficulties in learning, particularly in the field of spelling:

1 We found out that those children who had not yet mastered blending in our 6th examination (18th week at school) — i.e. they did not succeed in reading any of 4 synthesis-words (alt [old], Motor [motor], laut [loud], Hammer [hammer], or dropped off again in the following weeks — had difficulties in spelling later.

2 We found that retarded ability to blend correlates with other types of weakness/deficiency. This syndrome can be characterised as follows: difficulties in acquiring the ability to blend reappear in the form of lasting difficulties in spelling. The pace in learning, therefore, indicates difficulties in learning when acquiring reading/writing skills. There is a very distinct correlation to segmentising, to the syntactical dimension of linguistic competence and to the ability to represent the phonetic structure of a word in writing (cf. Dehn 1985a).

Further results of our examinations:

- All poor readers are weak in the phonetic synthesis (S-o-f-a). Good readers, too, are sometimes weak in the phonetic synthesis. This suggests that they read "in a different way" and do not depend exclusively on the phonetic synthesis (or analysis).

- Those children who succeed in blending at a late stage only, show in the course of the scholastic year (at least for one section) decreased achievement; a discontinuity in the process of learning.

When comparing the two groups of 79/80 and 81/82, the development of the ability to blend, we found an astonishing correspondence in the development of blending.(Dehn 1984, p.105). This is even more remarkable since in the two groups the importance of blending has been rated quite differently. This shows that — like knowing the graphemes — teaching and learning need not be analogous processes at all. On the one hand it seems that the ability to blend depends on the cognitive development of the pupil; on the other hand, the late mastering of this skill indicates some long-lasting difficulties in spelling, and therefore this is not merely a retardation in development that can be compensated for in every case in the course of the following years at school. These indicators that we found here have been confirmed by Peter May's study (1986). However, the prognostic value of these findings still has to be verified by tests with a larger sample.

(c) *The sequence of operations in the process of reading: the level of "meta-operations"*

We know from recent research that not only the ability to synthesise/blend is important for learning to read, but primarily its integration into all other operations is connected with the ability to structure and segmentise words or text. Up to now the operations have been investigated with regard to the given text and therefore, in particular, the motivation of reading errors.[1]

Due to our interest in the process of reading, we have researched our material in detail on how far, and in what ways, successful and unsuccessful ways of reading isolated as well as contextual words differ from each other. In our attempt to determine distinctive features of these successful or unsuccessful ways of reading we do not refer to any specific reading processes; since this is not a matter of just determining features of previously selected (operationally defined) pupils, a method that for good reason has been broadly criticized in the discussion of dyslexia. We have analysed the transcriptions on qualitative features, i.e. on the questions:

- how the child is approaching the task of reading a text;
- which steps he is taking towards the solution;
- how far some pupils, in spite of great efforts, still miss the solution or only reach it with considerable help of the text administrator.

We have found the following steps of reading (examples from the 18th week of schooling):

1 successive synthesis ⟶ part of the word or preliminary draft of the word ["Wortvorgestalt"] ⟶ word*

   e.g. h-a-t hat (successful)
        au a:l:'t alt (successful)

   e.g. *laut:* l-a-u-t ut/au Auto ault Auto (failing)
        laut l-a-u la:lt alt/l lar fula la lauf (failing)

---
[1] Gibson/Levin (1980); Scheerer-Neymann (1978), Goodman (1976), Brugelman/Fischer (1977).

93

2   Consonant-Vowel-Cluster ⟶ preliminary draft of the word ⟶ word

     e.g. la la lau: laut (successful)
           Ma Mo:'tor Motor (successful)
           Motor: Ma M-mau Maut/M:aut/ein o Mau (failing)
           hat: he h Hammer ha't at alt (failing)

3   draft ⟶ control (as a successive synthesis or as a CV-cluster) ⟶ word

     e.g. Mario M-a-r ja Mario (successful)
           lahm le:leise (successful)

     e.g. *Motor:* Mario/Ma/mat 'tor (failing)
           Motor: Mario/M-o-t-o-r mag Mond (failing)

4   draft ⟶ (probably control) ⟶ word

     e.g. halt hat (successful)
           Mari Motor (successful)

     e.g. *Motor:* Mann Frau (failing)
           laut: lau:f la:t lat (failing)

5   Prompt reading of the correct word/word not read: word misread without any trials or corrections.

1.7.3.4 Conclusion

The substantial result of our study is in our view: the different forms of sequences in the process of reading are in their total structure very much like the different forms of problem-solving thinking (Duncker 1962, Dorner, 1984): From a starting point (the graphic display of a word or a text) the beginning reader has to attain a goal that is not yet known to him (the comprehension of the word or context). In order to get there, the child has to carry out different partial steps (e.g. letter — sound correspondence, generating a string of sounds, structuring a string of letters, paying attention to the context and understanding what is written). These partial steps have to be co-ordinated towards the goal as well as controlled by considering the starting point. However, the child needs courage to do so, namely, to try out partial steps and drafts of the word at all, and he also needs accuracy and something like self-criticism when checking his own trials.

---

* Key of symbols:
-   = no auditory blending (no synthesis)
   = break, but no interruption of flow of articulation
:   = lengthening of sounds
/   = help of the test administrator

In detail we have found:

1 Good and poor beginning readers do not differ from each other in such a way that one group operates with a more holistic, the other with a more synthetic approach. Both strategies, the "top-bottom" as well as the "bottom-top" may succeed or fail. Nor do good and poor readers differ from each other in making reading mistakes or not.

2 However, good and poor beginning readers do indeed differ from each other in the *quality* of their mistakes. While the first are able to self-correct and to approach the solution step-by-step, the latter deviate more and more from the goal. Their strategy reveals breaches. They do not operate in a *stringent* manner. The word these children say as the result of a partial step barely has any similarity with what they ahve worked out before. There is no clearing up of word confusions; on the contrary these increase in the process of reading. Non-words (e.g. Mat'tor, lat) are also very striking in the group of poor readers.

3 The *consonant-vowel* cluster plays a special part in the process of reading (cf.2). It is an important indicator of progress in the process of learning; not only is it fundamental to blending but also to structuring, another important operation. Especially during the stage of acquisition, the syllable as a unit of articulation is of crucial function. It is a striking fact that children who were taught by means of a reading programme that (especially in the beginning) does not particularly stress blending as a teaching method, and that also allows holistic processes, against all expectations use at both times of testing more than twice as often a structuring approach to part of the word (cf. (2)) than children who were taught with a programme that emphasizes synthesis from the beginning on. Until the end of Grade 1 this way of reading increases. At the same time there is, however, a decrease of the holistic approaches to words in both reading programmes (cf. (4)).

4 Another difference between good and poor readers is the way they use the hints and support of the test administrator. Poor beginning readers are often unable to use these. We have characterized this as a *lack of flexibility.* This lack of flexibility is also indicated by the beginning reader's restricting himself to only a single strategy and his inability to consider the different qualities of words when operating. That is the case, for instance, when the reader troubles himself to blend an often taught word of the primer.

5 Hints of the test administrator are helpful in barely half of the interventions (cf Dehn 1984, p.112). This may be the case because the student test administrators were not trained. However, in a follow-up study that is geared to the development of systematic observation technics of learning ("LERNBEOB-ACHTUNG") and in which teachers are exclusively the participants, this result has been verified.

1.7.4 Further studies

In a theoretical as well as empirical study, Peter May (1986) scrutinized the acquisition process of literacy by means of categories of problem-solving. Here our findings have been verified and considerably developed. A central idea in May's work is that of a "heuristic competence" ("heuristische Kompetenz", Dorner,1979: 84), i.e. the learner's self-assessment of his abilities. For this is the driving power of his willingness and motivation for learning. That means the primary reason for the poor beginning reader's slow progress in the interaction with written language can be found in his way of approaching a task, for instance, when reading a text. His action has not enough initiative and confidence to increase his abilities in the process of reading; rather he falls back — often unnoticed — on strategies of avoidance and compensation.

The presented interpretation of our observations on learning to read from a cognitive point of view is closely related to the following questions:

- what kind of controlling and correcting processes does the first grader already have at his disposal?

- Which of them could be made analytically accessible to him, e.g. how do language awareness and acquisition of written language require and influence each other in the various dimensions of acting and verbal reflection?

This issue has been investigated in detail by Petra Hutlis (cf. 1985). In her study that is being analyzed at the moment, Petra Hutlis confronts beginning readers with readings of a hand puppet, thus giving them the opportunity of commenting and correcting. In this way even the poor reader is offered the part of someone who knows how to do it.

1.7.5 Consequences for instruction

The various ways of reading we have found are, (with the exception of successive blending), are seldom a topic of instruction. Our results indicate that a highly complex process such as the acquisition of written language develops something like private or individual principles, especially in regard to the significance of the material level and the development of individual operations and meta- operations. Our observations question the widespread idea of the precedence and dominance of teaching versus learning. Of course learning at school is directed learning; however, the child's way of learning is neither in a quantitative nor in a qualitative respect primarily determined by the method of instruction. That does not mean a limitation of the teacher's significance. Even the reading errors demonstrate that the children are in need of assistance; however, not in the manner of direct instruction through demonstration, but rather as a "following" support that considers the strategies and difficulties of the learner. In doing this the teacher ought to call his attention especially to the "meta-operations".

Instead of emphasizing the teaching of graphemes (material level) or of single operations (level of partial operations) we, as primary school teachers, should think about ways of making the sequence of steps in the process of reading a topic of instruction (the level of "meta-operations"). According to our findings, the function of knowing the graphemes is generally over-estimated, though it is, in its relation to the partial operations and the meta-operations, an essential component of the learning process, and blending as a technique is not readily accessible to teaching.

If we want to make the process of reading a topic of instruction, we need to provide opportunities to practise and experience this process, probably in small groups, and to talk about such strategies ("How do you know that you read the word properly?" "What do you do when you cannot read the word at once?" "Which part of the word do you know?" — cf. Downing's term of "cognitive clarity", 1972). This implies, again for the teaching of reading, a new way of handling mistakes. Instead of having the child's mistake only corrected by someone else, as is common up to now, the way of correction has to become obvious. In such cases, mistakes are causing learning processes. There would be a training of self-control, and, moreover, a training of how to integrate the perceived into the expected and conceived.

As yet we do not have enough experience in supporting pupils in "ways of finding out" (Heurismen). The question "What haven't I tried yet?", can, in its plainness and easy way of learning, hardly be surpassed (cf Dorner 1982, p.147). Possibly we have reflected too little on reading as a "mental act" from a methodological standpoint. This is true in the first place for those beginning readers who have long-lasting difficulties in the acquisition of literacy, and has extensive consequences for conceptions of supporting techniques and remedial teaching.

1.7.6 LEARNING TO WRITE AS AN ACT OF SPEECH ANALYSIS

1.7.6.1 General

Our interest in children's writing of words that had not yet been treated in the course of the writing programme (e.g. "Please write 'lamp!'"), was in the first instance only directed to obtaining disclosure about the "cognitive map" (Neisser, 1976) of a word that a child has in the stage of elementary acquisition of written language possesses. We intended to relate the results to the child's strategy of reading new words and texts. Therefore, when selecting the words, we have paid much more attention to their phonetic structure (length of the word, sequence of consonants and vowels) and to their relationship to the reading and writing programme, than to any difficulties of spelling. On the contrary, elements and aspects of the word which are difficult for the beginning writer, we found out only in the analysis of the writings (cf. Dehn 1985a).

The decision on which elements of speech ought to be put down, is pre-supposing an act of speech analysis. Coulmas (1981) is proving his thesis "writing is speech analysis" only by comparing different systems of writing. However, his ideas are suggesting parallels to children's acquisition of literacy. There is an important point, though, in which the activity of the beginning writer is different to that of an inventor of writing. The first already encounters the norm of written language in his/her environment and in the writing programme at school. Thus, children's activity of speech analysis is not only directed at speech but also at written language. Since written language is in several respects not merely writing down speech, the claim to children's activity of speech analysis is particularly complex.

1.7.6.2 Categories for Describing the Act of Speech Analysis of the Beginning Writer

Children's writing

- is analysed as to whether it is recognisable as being governed by rules (not in every case the rules of orthography);
- is classified as to how far the phonetic structure of the word is represented;
- is structured as to which features of spoken and written language the child gives his attention to.

In all but one of the observed classes some of the children's spelling could neither be read nor understood. Such *diffuse spelling* (e.g. "LBED" instead of "wo", "Ltl" instead of "Lampe") can reveal some kind of mechanical knowledge of the letter. The children are indeed aware of the fact that writing consists of letters, but they do not know their function. Therefore, they have a very hard time in finding an appropriate approach to literacy.

The spelling that is recognised as being governed by rules differs according to the degree of completeness. Some is *rudimentary* (e.g. "HDL" instead of "Hundeleine"). At first sight it may like like diffuse spelling, especially because of its shortness. However, the decisive difference is in that all letters (except perhaps one or two graphemes) have phonetic equivalents. Compared with the pre-school forms of writing like scribblings and paintings, the matching of single letters to phonetic elements of a word is a very important step in learning. It presupposes that lingual phenomena, that up to this point were unconsciously perceived in interaction, now become a matter of analytical awareness.

Other forms can be understood as developments of rudimentary spelling, but not necessarily as a step in a linear development towards acquisition of orthography: much of the spelling contains aspects of speech, which are irrelevant for orthography, e.g. the marking of the position of articulation (e.g. "BUR" instead of "Buch"), or peculiarities of drawing (e.g. "redeher" instead of "Rader"), or characteristics of dialect or colloquial speech (e.g. "Toa" instead of "Tor" in Hamburg).

Many children follow in their writing the phonemic principle of orthography. In many cases that leads to correct spelling, but in those cases, where graphemes are ambiguous, this leads to misspellings (e.g. "Kop", "Bal", "Reder").

From the beginning on, the children did not only orientate themselves to phonemes, but also to spelling, to "elements of orthography". The beginning writers noticed that sometimes two of the same letters follow each other (e.g. "lammpe"), that sometimes there is "ck" (e.g. "Schranck"), that sometimes there is "er", where "a" is heard (e.g. "Sofer" instead of "sofa"), and they generalise their findings to the wrong subjects. It seems as if children at first simply test graphemes, which go beyond the plain sound – letter – correspondence. Later on they acquire the specific marks of the spelling of a word and master the morphematic principle of orthography as well.

1.7.6.3 Individual ways of learning and forms of processes of spelling

The way in which the child acquires spelling changes during the first year at school. Most of the children succeed in considering the chronological order of the phonetic structure of the word. Compared with that they have much more difficulty in carrying out a complete phonetic analysis, or in attending only to the distinctive function of phonemes of standard articulation. Occasionally children use "elements of orthography" even during the first few weeks, but usually that appears more often in the second half of the first year.

The *ways of learning* of the children we observed differ from each other in regard to the starting point of the activities of speech analysis, and in regard to the degree and quality of children's approach to literacy during the first grade. No way of learning is like another, but nevertheless the 66 ways of learning we have studied can be classified as follows. We found four different ways of learning:

- Ten out of 66 children were able to follow the reliable way of *phonemic spelling from the very beginning.* They could already analyse the phonetic structure and start to come to terms with spelling at an early stage. None of these children had long-term spelling difficulties during the first four grades.

- Thirteen out of 66 children were at *first engaged in representing their own articulation.* But very soon they also got confidence in dealing with the principles of spelling.

- Many of the children (27 out of 66) were engaged for a long time in breaking up the phonetic structure. They *started with rudimentary spelling,* which they developed increasingly. Their learning closely corresponded to our ideas of spelling acquisition.

- At any rate 16 out of 66 children had either *difficulties in basic orientation to spelling or their acquisition process stuck to mainly rudimentary spelling.* Difficulties in basic orientation not only became obvious in diffuse or rudimentary spelling; these children refused to write much more often than the others.

EXAMPLES	8th week Tor / los	14th week Marmelade / Limonade	25th week wunderbar	36th week Tulpe	Kinder
(1) Peter	Tor	MARMELADE	wUNderbAr	Tulpe	Kinder
Kerstin	Tor	MaMELRAde	wun BA	TulPe	Ki der
(2) Sven	Los	LMonAt	fontaPa	tolqe	Kender
Meike	ols	liMonade	womerba	Tolbe	kinder
(3) Thomas	Lo	Li o q	fo q q	Tolpe	Kender
Silke	T	M ma Lot	Won pra	Tulpe	Kinper
(4) Anja	T	W t	MonePA	tolPe	KI R
Sönke	—	M W	M d M	la P	k n r

An important result of our long-term study is that all children who in primary
school have enduring difficulties with writing belong to the last group. (But the
inverse is not true. Some of the children with serious difficulties at the beginning
acquire a productive way of dealing with literacy.).

The pressing question arises whether we could discover such basic difficulties
children show in spelling untrained words much earlier, if we give them the
opportunity. Perhaps they would be saved from much failure by intensive writing
lessons in class, or by individual help.

1.7.6.4 Consequences for instruction

1 If the process of writing acquisition is governed by rules, as our studies have
shown, the syllabus, i.e. the vocabulary of reading and writing, has to represent
the most important principles of orthography from the very beginning. That
means, to practise from the beginning short words that contain "elements of
orthography" (e.g. "Ball", "Hund"), as well as words whose spelling follows
the phonemic principle. Later on simple words have to be chosen, whose
spelling contains difficulties according to dialect or colloquial language (e.g.
Turm, Kinder, Leiter for children from northern Germany).

2 If an early secure base is useful for writing acquisition, the vocabulary should
be small and should vary according to the level of achievement. In this way
poor writers can gain confidence, too.

3 When children's speech analysis in learning to write is not random but governed by rules, the children should be encouraged to write "spontaneously" in the first stages of learning, i.e. instruction should not be restricted to copying, which is still common in normal school practice. For, especially in spontaneous writing, children can experience for themselves the problems of spelling and recognise those problems. That does not mean, though, that they can work out the regularities of orthography for themselves. At this point instruction and guidance is needed, and probably the acquisition of patterns as well. This second consequence only seems to be in contrast to the former, for learning and teaching are not simply corresponding processes. It is not contradictory to gain on the one hand a secure base with the clear subjects of the teaching programme, and, on the other, to experiment with rules and to develop them while writing spontaneously. And, through the latter, the teacher can get an idea of his pupils' difficulties and information on the "sensible phases" of their learning processes, if he knows how to read their mistakes.

Teaching and learning to write would then again be fascinating intellectual activities.

1.7.7 CURRENT STUDIES AND OPEN QUESTIONS

On the basis of these results we have, in recent years, with the co-operation of the school administration in Hamburg and with financial support from the Deutsche Forschungsgemeinschaft, developed and evaluated an instrument for the systematic observation of learning processes ("LERNBEOBACHTUNG"). It consists of reading and writing tasks in three terms of the first grade: November, January and May, and is independent of the specific teaching programme. The aim of "LERNBEOBACHTUNG" is to understand the child's approach to learning to write and to prevent difficulties in the basic orientation to literacy at an early stage. Its standard is primarily the individual learning process and only secondly the general norm.

In the last few years I have extended the observations to older pupils of primary school and pupils at special schools, as well as to adult functional illiterates. I have also integrated "narrative interviews" of the latter's school history. Furthermore, we have made a set of 80 video scenes from remedial and small group teaching ("Forderunterricht") at primary school, in which the teacher helps a child with an actual difficulty in reading a word. The aim is to find out and define conditions for successful help and hints. Already at this point it is becoming obvious that it is extraordinarily difficult, especially with poor beginning readers and writers, to initiate acquisition of literacy as an active and self-contained process of forming rules and to avoid discrimination and discouragement. Poor learning is often based, it seems, on refusal and defence, i.e. blockades by the learners and, sometimes, by the teachers (cf Heuser 1985).

At the moment the problem of illiteracy is attracting wide public interest in our country. In my opinion, it is an urgent task of research to find out the causes in elementary schooling, in order to avoid the serious consequences for the personal and professional development of those concerned. For the interpretation of the learning processes and the results of "LERNBEOBACHTUNG", I prefer the concept of "cognitive schemes" (as far as it is possible), to that of "perception-disorders", that has recently been discussed again. (cf Schenk-Danziger 1984).

REFERENCES

Balhorn H (1983) Rechtschreibenlernen als Regelbildung. Wie machen sich Schreiber ihr ortografisches Wissen bewusst? In: Dis.Dt. 14, S. 581-595.

Balhorn H and Vieluf U (1985) Fehleranalysen-ortografisch. In: Disk.Dt. 16, S. 52-69.

Bauer S (1971) Die Fibel als Instrument der Sozialisation. In: Disk.Dt.5, S. 265 ff.

Bergk M (1980) Leselernprozess und Erstlesewerke. Analyse des Schriftspracherwerbs und seiner Behinderungen mit Kategorien der Aneignungstheorie, Bochum.

Bosch B (1984) Grundlagen des Erstleseunterrichts, Ratingen 5. Aufl. 1986 (1937) Reprint.

Brugelmann H and Fischer D (1977) Verlesungen — Fehler oder (diagnostische) Hilfen? In: D Grunds. 9 S. 575 ff.

Brugelmann H (1984) Lesen – und Schreibenlernen als Denkentwicklung. In: Z.f.P 30 1, S.69 ff.

Castrup K H (1978) Spontanschreiben zum Erwerb der Schriftsprache. In: D Grunds 10 S.445 ff.

Castrup K H (1984) Das Aufschreiben im 1. und 2. Schuljahr. In: Bayerische Schule 21, S. XXff.

Coulmas F (1981) Über Schrift, Frankfurt.

Dathe G (1974) Einführung in die Methodik des Erstleseunterrichts, Berlin (Ost).

Dehn, M (1975) Texte in Fibeln und ihre Funktion für das Lernen, Kronberg.

Dehn M (1977a) Text und Übung im Leselehrgang. In: Schwartz E (Hg.): Lesenlernen — das Lesen lehren. Frankfurt 1977, S. 86ff.

Dehn M (1977b) Phonologie und Erstleseunterricht. In: M Grunds 9, S282 ff.

Dehn M (1978) Strategien beim Erwerb der Schriftsprache. In: D Grunds. 10 S. 308ff.

Dehn M (1984) Lernschwierigkeiten beim Schriftspracherwerb. Kriterien zur Analyse des Leslernprozesses und zur Differenzierung von Lernschwierigkeiten. In* Z.f.P 30, 1, S.93 ff.

Dehn M (1985a) Über die sprachanalytische Tätigkeit des Kindes beim Schreibenlernen. In: Disk. Dt. 15, 81, S.25-51.

Dehn M (1985b) Wortschätze für Schreibanfänger. In: Grundschule 17 (1985) S.21 ff.

Dehn M (1985c) Zur Beobachtung von Lernprozessen und Lernschwierigkeiten in Lese- und Schreibkursen. In: GIESE H (Hg.): Die wissenschaftliche Fortbildung von Kursleitern in der Alphabetisierungsarbeit. Universität Oldenburg 1985, S. 14ff.

Dehn M (1986a) Über die Aneignung des phonematischen Prinzips der Orthographie beim Schriftspracherwerb. In: Brugelmann H (Hg.): ABC und Schriftsprache: Rätsel für Kinder, Lehrer und Forscher. Konstanz 1986 S.97 ff.

Dehn M (1986b) Lese- und Schreibschwierigkeiten verstehen lernen. Projekt mit Schulanfängern und erwachsenen Analphabeten. In: Grundschule 18 (1986) Heft 3 S. 18f.

Dehn M (forthcoming) Lernbeobachtung, Lesen und Schreiben in Klasse 1 als Voraussetzung für frühe Lernhilfen.

Dorner D (1979) Problemlösen als Informationsverarbeitung, Stuttgart 2.

Dorner D (1984) Denken, Problemlösen und Intelligenz. In: Psych.Rsch.35, 1, S. 10ff.

Downing J (1972) The cognitive clarity theory of learning to read. In: Southgate, V (ed): Literacy at all levels. London, S.63 ff.

Downing J and Valtin R (ed) (1984): Language awareness and learning to read, New York, Berlin.

Drecoll F and Muller U (Hg) (1981) Für ein Recht auf Lesen. Analphabetismus in der Bundesrepublik Deutschland, Frankfurt.

Ferdinand W (1970) Über Erfolge des ganzheitlichen und des synthetischen Lese-(Schreib)unterrichts in der Grundschule, Essen.

Geiss M (1972) Die Konservierung sozialer Rollen. In: Doderer K (Hg.): Bilderbuch und Fibel, Weinheim.

Gibson E H and Levin H (1980) Die Psychologie des Lesens, Stuttgart (amer.1975).

Giese H W and Glass B (1984) Analphabetismus und Schriftkultur in entwickelten Gesellschaften.

Das Beispiel der Bundesrepublik Deutschland. In: D. Dtu 36, H.6, S. 25ff.

Goodman K S (1976) Die psycholinguistische Natur des Leseprozesses. In: Hofer A (Hg.) 1976 S.139 ff.

Goodman K S Analyse von unerwarteten Reaktionen beim oralen Lesen. In: Hofer A (Hg.) 1976 S.298 ff.

Gromminger A (1970) Die deutschen Fibeln der Gegenwart, Weinheim.

Hannig C and Hannig J (1974) Der Einfluss des Erstleseunterrichts auf die Sprache von Schulanfängern. In: C Hannig (Hg.) (1974): Zur Sprache des Kindes im Grundschulalter, Kronberg, S. 98ff.

Heuser A (1985) Probleme des Lehrens und Lernens in der Alphabetisierung — eine Fallstudie. Unveröffentl. Examensarbeit Hamburg.

Hofer A (1976) Lesenlernen: Theorie und Unterricht, Düsseldorf.

Huttis P (1981) Zur Funktion des Segmentierens beim Lesenlernen. Unveröffentl. Examensarbeit, Hamburg.

Huttis P (1985) Tobi mach Fehler — was nun? In: Grundschule 17, H.10.

Leont'ev A A (1975) Psycholinguistische Einheiten und die Erzeugung sprachlicher Äusserungen, Berlin (russ. 1969).

May P (1986) Schriftaneignung als Problemlösen. Frankfurt v.a. (P Lang).

Menzel W (1981) Schreiben-Lesen. Für einen handlungsorientierten Erstunterricht. In: Neuhaus-Siemon E (Hg.) (1981) S.134 ff.

Merkelbach V (1973) Lerninhalte in neueren Fibeln. In: Disk.Dt. 12 1973, S. 103ff.

Muller R (1979) Leseschwäche, Leseversagen — Legasthenie. 2 Bd., Weinheim.

Neisser U (1979) Kognition und Wirklichkeit, Stuttgart (amer. 1976).

Scheerer-Neumann G (1978) Die Ausnutzung sprachlicher Redundanz bei leseschwachen Kindern. In: Zt. für Entwicklungs- und Päd. Psych. 1 S. 35 ff.

Schenk-Danzinger L (1984) Legasthenie. München.

Schlee J (1976) Legastenieforschung am Ende? München.

Schmalohr E (1971) Psychologie des Erstlese- und Schreibunterrichts, München.

Sirch K (1975) Der Unfug mit der Legasthenie, Stuttgart.

Spitta G (Hg.) (1977) Legasthenie gibt es nicht - was nun? Kronberg.

Wiegl E (1974) Zur Schriftsprache und ihrem Erwerb. In: Eichler S and Hofer A (Hg.) (1974) Spracherwerb und linguistische Theorien, München. S.94ff.

Weinert F u.a. (1966) Schreiblehrmethode und Schreibentwicklung, Weinheim.

Weinert F E and Kluwe R H (Hg) (1984) Metakognition, Motivation und Lernen, Stuttgart.

Wode F (1978) Fehler, Fehleranalyse und Fehlerbenötung im Lichte des natürlichen lingua-2-Erwebs. In: Ling.u.Did. 9, S.233 ff.

1.8 A ROUGH OUTLINE
OF THE SITUATION IN SWEDEN

by

Birgita ALLARD and Bo SUNDBLAD, Sweden

1.8.1 SUMMARY

A brief overview of developments in the teaching of reading and writing in Sweden is given. While the situation is a dynamic one, concern is expressed about the lack of knowledge of the theory on which teaching methods are based.

1.8.2 TEACHING READING

1.8.2.1 General

Before the seventies, methods used for teaching children to read were never called in question and synthetic methods had, up to then, dominated. In the 1950s the alphabetic method was replaced by phonics. Children had to start with the smallest units — letters/sounds — and learn them one by one and how to sound them together into words. Reading primers were introduced, based on this synthetic/phonic method. In order to help the children to realise the idea of "sounding together" these units, they were given the following illustration:

1960

1986

Later on, reading primers were introduced based on an analytic method, starting with whole words, which had to be analysed into their separate sound/letter units. The differences between the two approaches are minimal. Today all the reading primers (with one exception) are quite alike, whether they are synthetic, analytic or a combination of the two, with a heavy stress on phonics.

1.8.2.2 The Learning Experience Approach

In the seventies however, a Swedish version of LEA (Learning Experience Approach) was introduced. No reading primers are used in this method. The teacher starts out from the children's own utterances about things they want to communicate. The teacher writes them down and then the children can "read" them. Each child can choose a word which he or she likes in order to analyse it into its parts. This approach is analytic, starting with meaningful sentences, passing on to words down to the sound/letter units. Learning sound/letter units and their combinations is central also in this approach.

Widespread interest in LEA — many teachers began to practise it — resulted in an intensive discussion about methods for reading instruction — a debate which is still going on. It also initiated further research on reading, which had, in this area, hitherto been limited either to reading disabilities or to test production. Since the debate was sometimes heated, especially that concerning the effects of the different methods, research for evaluating the different approaches was initiated and carried out. But that gave rise to yet another discussion, about the procedures for measuring the effects of such different methods. Teachers working with LEA maintained that the traditional testing programmes did not give justice to "their" children's reading ability, since they, from the start, were taught to read for meaning and not primarily for decoding lists of words. This topic grew into a general discussion about the use and value of psychometric tests in school at all levels. Although the discussion initially focussed on the basic process of learning to read, it soon extended to literacy development in general, including the intermediate and senior levels, initiating research in this area too.

1.8.2.3 The synthetic – phonic approach

At the same time a "new" method became popular. It was said to be the best method for children and teenagers with serious reading difficulties, since the method means re-learning from the beginning. This method can be described as the most consistent and logical variant of the synthetic-phonic approach. The children start with learning the vowels and then one consonant at a time. Before going to the next consonant the first is combined with each vowel and that combination of consonant/vowels as to their sounds has to be learnt and automised through over-learning. Reading during the first year in school is concentrated on the decoding of nonsense syllables of two, three and later four letters.

1.8.3 CURRENT DEVELOPMENTS IN TEACHING READING

These methods are still current, although LEA has lost some of its ground. But interest in reading at all levels is still growing, mainly because of the alarming reports of large numbers of children leaving school with inadequate reading ability. While trying to find the causes of this complex situation, different aspects have come into focus, such as the current research on dyslexia; pre-school children's interest in reading and writing; scrutiny of all the texts that school children are forced to read year after year. The latter aspect is for the moment of immense interest. The character of textbooks on different subjects had changed radically during the past thirty years. The dividing line can be located to the middle of the sixties. Before that, texts were written in a personal and engaging way, with examples relating to the children's experiences. The texts were extensive with few but informative and instructive drawings and pictures, with no or few supplementary books for exercises. Nowadays, the pupils are faced with a set of books purporting to be self-instructive: one textbook containing information and facts, one, two or more booklets for training and exercises, with "semi-manufactured" tasks where they have to fill in a missing word or a short answer to a question. Generally the texts are short, impersonal and factual, containing many coloured pictures with brief captions. Every sentence transmits two or three items of information. Reducing a text from, let us say, ten pages to one, still including the same number of facts or even more, necessarily means much higher levels of abstraction. A ten-year-old has to read the following: "Industries in Turin get their energy from ...". But if one asks a child what he understands by the word "industries" he or she may answer "grey buildings". If the author, instead of using only four words had used 23 like this: "In Turin there are factories where they make cars. You have certainly heard of, and seen those cars of the make of Fiat", then every child would have a chance to understand. From the moment they can read (7-8 year old) the children are confronted with texts which are impossible for them to comprehend in the same way as the authors want them to understand. What does this mean for their literacy development? (And of course for their knowledge in different subjects?).

When they start school many children can already read. Those who are strict adherents of the synthetic/phonic method claim that this is bad, because the children may have learnt to read in the wrong way and must therefore re-learn ie go back to the starting point once again. They argue that the best thing to do is to prevent children from learning to read before they start school. Another argument in favour of this is that if a child can already read, he or she will become bored during the first year in school. But the opposite view has become more common. It is obvious that many children become interested in written language long before they start school. Encouraging and stimulating this interest, answering the children's questions and above all helping and showing them the communicative function of written language is a very common strategy.

During recent years a new "trend" has appeared, with relevance mainly for "reading readiness" or preparatory training and for children with reading and writing disabilities. It is a purely neurological approach where reading is defined as a set of sensorimotor functions. Training of all these sensorimotor functions and of their co-ordination will facilitate all learning, and especially learning to read and write. According to this view, the causes of reading and writing disabilities are to be found in the sensorimotor functions. These ideas have spread like wildfire, especially at the pre-school and the primary levels.

1.8.4 WRITING

Interest in writing is only quite recent. Up to then, writing has been of secondary importance and a question of learning to form letters correctly, to spell, and so on. However, influences mainly from writing projects in the USA are slowly changing this view of writing and this is sure to initiate interesting research projects for the coming years.

1.8.5 TEACHER TRAINING

To bring about a far-reaching change in the teaching methods practised in schools depends very much on the quality and quantity of in-service training. In 1980, Sweden had a reform of in-service education. The general aim of this reorganisation was to enable this training to be directed more closely towards the realisation of the goals in the curriculum, and at the same time to lower the cost and time per person in order to reach a greater number of teachers. The arrangements differ from place to place and from short courses (one week) for many teachers with hardly any results to longer courses (five to ten weeks) for fewer teachers but with more effective results. Thanks to the current intensive debate on reading, many courses have centred on this topic and thus an awareness of reading and its dimensions has become widespread among teachers, although knowledge about writing is still in embryo.

Teacher training is also of importance, although as a whole it has not changed very much since the sixties. However, reorganisation of teacher training has been decided upon by the government and will come into effect in 1988. It is too early to predict the consequences of this reform for the direction of research on, and teaching methods in, reading and writing.

A general problem, which is relevant both for in-service and for teacher training, is, however, the non-existence of knowledge of the theoretical basis of the methods taught and practised. In order to be that institution of higher education which it is supposed to be, teacher training, as well as in-service training, must include a theoretical competence and thus establish a connection with different research approaches.

To sum up: the situation in Sweden is, to say the least, dynamic.

1.9 REPORTS OF THE WORKING GROUPS

1.9.1 REPORT OF GROUP A

Chairman: Mr Gérard CHAUVEAU
Rapporteur: Mme Dominique LAFONTAINE

In view of the term "new" or "functional" illiteracy, it must be a priority concern to highlight possible causes of this and to examine *strategies of improving literacy skills.*

For many pupils the teaching and learning of literacy does not extend beyond the first two years of education, thus the extended skills or reading for a variety of purposes is underemphasised.

A possible solution to this may be that, from the beginning of school education, the teaching of reading and writing should be integrated through strategies which have meaning for the child and lead to an understanding of the functions of literacy.

The child's understanding and perceptions of the nature and purpose of written language has been found to predict later success in reading and writing. For many, however, these perceptions relate to mechanical or institutional aspects; to learn new words or to avoid school failure. Personal motivation for reading and writing is low in such groups.

Strategies which may improve levels of literacy might include:

- building on children's oral language as a transition to reading and writing (*thus accounting for cultural and linguistic diversity*);
- allowing more space and time for children to be engaged in extended reading and writing activities;
- teachers themselves providing a model, both of a reader and writer, to help children understand the processes involved;
- accounting for the individual needs of learners and their levels of development;
- emphasising *textual understanding* as a priority in reading and through this *systematically* developing the *skills of decoding.*

The interactive model of reading takes account of both decoding and comprehension skills, according to the needs of the learner. It must be emphasised that all readers need a range of strategies which will help them develop a flexible approach to reading texts.

Research findings suggest that the same teaching methods may produce different results in individual children's reading achievement. This emphasises the *importance of understanding* and *diagnosing individual needs.*

In *writing* vital areas of consideration are that children:

- are aware of the purpose of writing;
- realise that text varies according to the message and the audience. Identification of an audience is a motivating factor in adopting the conventions of writing, spelling and punctuation;
- recognise the higher order skills of planning and composing as more important in the early stages than "correctness";
- collaborate during the writing process with teachers and peers, and are able to plan and revise texts;
- see writing in relation to other aspects of the curriculum, mathematics, aesthetics... etc.

Crucial "out-of-school factors" in the growth of literacy are identified as:

Community involvement: partners in education to include parents, librarians, ... etc.

Easy access to books which may help to develop all children's awareness of print and its functions.

Greater support and *resource input* to develop interest in literacy in defined "underprivileged areas".

Initial and inservice training of teachers must be a priority. This should ideally be school-based and consist of groups of teachers working with lecturers and researchers. From such innovations, teachers themselves may become disseminators of ideas and information.

It is vital that communication between teachers, teacher trainers, education inspectors and researchers is established in order to develop effective policies which will raise the levels of literacy.

1.9.2 REPORT OF GROUP B

Chairman: Dr Sjaak SANDBERGEN
Rapporteur: Prof. Michael STUBBS

The term "the new illiteracy" is widespread, but potentially very confusing rather than clarifying, since it is not itself a fact, but an *interpretation.* It is a term which is situated within a particular political and educational discourse, which speaks also of "falling standards" of literacy, and which sees linguistic diversity as a "problem" rather than as a resource. If standards *are* falling (and the evidence is very difficult to interpret), this may be because demands are rising. It is probably more serious to be illiterate nowadays in highly bureaucratic, institutionalised, socially stratified societies.

Expectations of appropriate literacy differ greatly across different social groups. And demands on *writing,* as opposed to reading, are increasingly important, if people are to have access to those who hold institutional power.

In a situation in which the demands on literacy are increasingly complex, it is increasingly important that teachers understand the theoretical principles underlying literacy and literacy teaching, and the principles by which children actively acquire spoken and written language. It is particularly important that they understand that reading and writing have very different forms and functions in different communities. They must understand a great deal about how language works: how writing systems are organised; how spoken and written language differ; how people interpret meaningful language in context. Without such understanding, teachers are condemned to mechanical and not flexible teaching methods. The development of this understanding raises problems of teacher-training, but is based on the belief that it *is* possible to formulate an appropriate educational theory of language.

Much of the debate on teaching methods hinges on distinctions between decoding versus comprehension of text, or accuracy in spelling versus creativity in writing. It is often claimed that these oppositions represent unproblematic choices between techniques of teaching. In reality, the choices are complex and deeply symbolic, because they relate to several other oppositions which are highly value-loaded: tradition/innovation, teacher-centred/child centred, artificial/natural, passive/active.

The choices relate, in other words, to a theory of children. This point implies, in fact, that the oppositions must not be assumed and taken for granted: the important issue is the relationship between the pairs in each opposition.

These points are preliminary, but essential, since debates in this area tend to crystallise deep-rooted cultural conflicts and are, therefore, difficult to maintain at a purely rational level. Everyone holds *theories* about literacy, even if they claim not to.

With regard to classroom practice, it is essential to distinguish *means* from *ends*. Ends might (or might not) include the ability to spell accurately. And one might debate whether this is an appropriate aim or demand for different stages in a child's schooling. Assuming one accepts the importance of this aim (given its enormous social and conventional significance), a quite separate debate concerns the means by which one might attain the aim: this might be via a teacher-centred and explicit teaching of rules, or it might be via a child-centred and implicit induction of the rules which relies on the child's own general linguistic competence.

In fact, the goals are rarely in dispute. It would be socially unrealistic to abandon the demand that children require a firm grasp of the conventions of standard written language. It is advisable, however, not to insist on this too early. Such an insistence can hinder children's own *use* of written language, and this use is the main way of learning the conventions. This implies that children must produce much more extended writing for conceptual and creative purposes for a real audience. If teachers have the necessary knowledge of the *principles* of language use and structure, they are in a strong position to organise teaching situations which can make use of children's own linguistic resources.

We return to the point that the essential issue involves the long-term goal of changing the point of view from which teachers regard children's own developing linguistic competence. The most important factor is how the teacher interacts with the children. The essential issue is, therefore, one of the principles underlying teacher-training, not of the relatively limited choice between competing methods of teaching literacy. This has obvious financial implications for educational departments and ministries in the member states.

1.9.3 REPORT OF GROUP C

Chairperson: Ms Birgita ALLARD
Rapporteur: Ms Edeltraut ROEBE

"The new illiteracy" cannot be described in exact figures, but by brief answers to the questions:

What is it?
How can it be described?
What's causing illiteracy?
What's to be done about it?
Why are we worrying about it?

Illiteracy as a term is relative, taking into account that reading is more than decoding: Reading is for acquiring new knowledge, for solving real-life problems and for pleasure; it is also the key to the world of the written language.

Reading demands are increasing in our society. Each individual is expected to be able to respond to the skills and attitudes required to manage public and private life in this society. Reading demands change as society develops and only the quantitative and qualitative growth will ensure that the standards are maintained.

It is generally accepted that functional illiteracy is not only restricted to the problems of analphabetism, dyslexia, etc. but also includes people having severe problems to master the demands for reading and writing in their everyday life. Illiteracy is caused by a number of interrelated factors among which the following can be mentioned:

- People with reading difficulties are not provided with sufficient opportunities where reading and writing could be experienced as leisure and pleasure, as enrichment of communicative means and as essential instruments to master everyday life.

- School may not be willing or able to deal with the children's different potentialities. A major reason for this is that school often wrongly assumes the mastery of some basic skills in reading and writing, accepted standards in social behaviour and readiness for learning and achieving when children start school.

- Children, poor in language, possibly due to their low socio-economic background or social and emotional deprivation, have problems in compensating for their deficient starting-point through and at school. Initial problems lead to learning problems, lack of self-confidence and self-esteem.

- Children labelled and accepted as poor readers may develop into illiterates because they will accept their position and eventually correspond with the picture in a kind of self-fulfilling prophecy.

- In many cases inadequate teaching contributes to illiteracy.
  + Aims and goals are often given lip-service and rarely realised.
  + Written material inside and outside school is not addressed to the "new illiterates".

- The learning atmosphere may also be a source of diminishing selfconfidence, demotivation and self-esteem and this might lead to illiteracy.

- The home, the environment, the school and society itself can be responsible for illiteracy because they often serve as inadequate models. Illiterates are often isolated, alone with their problems, anxiously hiding their disability, missing teacher's help. Society seems not to have enough time and understanding to support them.

In the light of the above, measures must be taken to ensure that illiteracy be kept to the barest minimum.

Unless governments realise that something must be done about it, the price that they will have to pay will surely be much higher than it will cost them to embark on projects to prevent it.

These projects can be aimed at:

- improving teacher training;
- organising inservice training;
- producing textbooks which children really like and which will make them like reading;
- "educating" parents;
- establishing research projects intended to identify problems of illiteracy and possible solutions.

PART 2:
NATIONAL REPORTS

2.1 READING COMPREHENSION IN PRIMARY EDUCATION IN FINLAND

by

Anneli VAHAPASSI
Institute for Educational Research, Jyväskylä

2.1.1 SUMMARY

In the Finnish comprehensive school curriculum, reading comprehension is considered the essential factor of reading skill and one of the fundamentals of learning. As a phenomenon, reading comprehension is understood to be an intricate interaction process which is influenced by the characteristics of both the text and the reader and by the reader's experiences. If comprehension proceeds successfully, the reader is able to utilize information contained in the text as well as extraneous knowledge in appropriate proportions.

This paper outlines the general process of reading comprehension into different levels. It is a synthesis of the conceptions of various researchers, and it distinguishes between the following levels of comprehension: reproducing, interpretative, evaluative and creative understanding. There are also suggestions for improving teaching and teaching materials.

2.1.2 STUDENTS' READING COMPREHENSION SKILLS AT THE PRIMARY LEVEL (1-6) OF THE FINNISH COMPREHENSIVE SCHOOL

2.1.2.1 General

Are primary school students able to understand different texts in a way which meets the requirements of the changing school and community? The investigation of this question has been carried out at the Institute for Educational Research since the early 1970s, most extensively through the *First National Survey of the Comprehensive School in 1979.* This nationwide assessment related the teaching practices and learning outcomes of the comprehensive school to the defined basic objectives and core subject matter. Among other things, the national survey sought to determine how students at the lower level of the comprehensive school master the following aspects of reading comprehension:

The 6th Grade (the end of the primary level)

C.1 Students are able to acquire and improve their knowledge by reading.
C.2 Students understand what they read.
C.2.1 They can intepret in their own words the meaning of words and sentences which they have read.
C.2.2 They can utilize the context of the text when inferring the meaning of words.
C.2.3 They are able to present in their own words the events and main points of the text.
C.2.4 They are able to draw conclusions from the content on the basis of hints contained in the text.
C.3 Students are capable of interpreting, in the above mentioned ways, fiction and informative texts consisting of normal sentence structure and dealing with relatively familiar topics.

The 3rd Grade

C.1 Students are capable of acquiring information by reading.
C.2 Students are able to read accurately texts with normal sentence structure dealing with very familiar topics.
C.3 Students are able to interpret in their own words the events and main points of the type of texts described above.
C.4 Students can draw simple conclusions from the events of such texts.

2.1.2.2 The investigation

The investigation used a method of measurement which considered reading comprehension to be a phenomenon which cannot be studied by means of a few tests and a small number of items. Therefore, in the national survey, the text samples were drawn from a larger set of texts in order to represent as much as possible in the types of text which students have to read in their community.

The test passages for the third and sixth grades were selected using criteria resulting from the author's earlier studies, which compared the reading comprehension skills of third and sixth graders with the difficulty level of learning materials. Texts which had been used in earlier studies were also included in the tests of all grade levels.

Levels of reading comprehension which were chosen as targets of measurement included recognising, interpretative and evaluative comprehension. For the measurement of the different levels 319 items were constructed for the ninth grade, 226 items for the sixth grade, and 172 items for the third grade. Of these items 66% in the sixth grade and 48% in the third grade were classified as items measuring basic objectives. The following table shows the solution percentages of the basic objective items in the different student groups. The items have been divided according to the text genre (fiction/fact).

Table 1. Average solution percentages and dispersions in grades 3, 6 and 9 in different student groups.

	3rd grade				6th grade				9th grade			
	Fiction items		Fact items		Fiction items		Fact items		Fiction items		Fact items	
	%	s	%	s	%	s	%	s	%	s	%	s
Total	72	18	71	14	76	18	74	19	79	18	69	20
Gender												
girls	76	18	74	15	77	19	77	19	81	19	68	22
boys	68	18	68	14	74	18	71	19	76	19	65	21
Population density												
town	75	18	74	14	74	23	75	19	79	19	66	21
rural community	70	18	68	15	75	18	73	19	78	18	67	22
School size												
small <80 (upper level <350)	72	20	69	16	77	17	75	19	80	19	69	21
big >80 (upper level >350)	72	17	71	13	75	19	73	19	78	18	66	21
Class type												
Single grade class	72	17	71	14	75	20	74	19	-	-	-	-
combined grade I	72	19	71	15	76	18	73	20	-	-	-	-
combined grade II	70	21	67	20	78	21	78	22	-	-	-	-
Teacher's rating of achievement of basic objectives in reading												
non-achievers	52	21	48	18	67	26	61	21	69	21	59	23
achievers	75	18	74	14	78	18	76	19	81	18	67	22
no rating	74	17	77	15	74	24	77	20	80	21	65	22
Ability groups in language (6th grade - intends to choose)												
basic course	-	-	-	-	64	22	59	23	68	21	57	22
intermediate course	-	-	-	-	74	18	72	21	76	20	63	23
long course	-	-	-	-	79	19	80	17	85	18	73	23

The table indicates that the basic objectives of reading comprehension on the primary level of the comprehensive school have been achieved by 70-80% of the students. Boys were somewhat less successful than girls in solving the same items. On the other hand, factors like rural area versus town or the size of school did not cause the kind of difference in solution percentages as gender.

Whether the result is considered to be satisfactory or not depends on the interpreter's point of view. We may think that a very good result has been achieved when over 70% of the total age group has attained the intended basic objectives. This view is supported by the finding that reading comprehension for the whole age cohort has somewhat improved during the 1970s, compared with the earlier binary school system. The skills of third graders had also improved a little between the years 1974 and 1979. At the same time, however, the standard deviation had increased.

On the other hand, the results can be considered alarming. For example, they indicate that approximately 20-30% of the students have not reached the skill level which is demanded of them by the text- books.

The national survey shows that students understand texts moderately well on the surface level, but they are not as successful in evaluative and deductive comprehension, as table 2 indicates.

Table 2. Average solution percentage of all reading comprehension items (x) and standard deviation (s).

	Reproducing comprehension	Evaluative and inter-pretative comprehension	N
3rd grade	x = 66% s = 20 number of items = 81	x = 52% s = 18 number of items = 91	1019
6th grade	x = 76% s = 15 number of items = 106	x = 63% s = 20 number of items = 120	1155
9th grade	x = 78% s = 15 number of items = 139	x = 65% s = 20 number of items = 180	1366

2.1.2.3 Possible explanations

The results of this survey support the general argument that schools emphasize the acquisition of the learning material as such, and its repetition as opposed to real understanding. Reading comprehension and learning stop at the surface level. Effective strategies for understanding have not been created. One of the reasons may be that students have not had opportunities for developing their own ideas, nor for discussing causes and effects, let alone for argument.

Another possible explanation is school practices. After the first two school years, a great leap takes place in school work. Until then children have been learning to read; now they have to learn many kinds of things through reading. At this stage students have not yet generally acquired efficient reading comprehension strategies. Often they are not able to realize what they understand of the reading matter and what they do not understand. In addition, they do not know what they should do if they do not understand.

In terms of reading comprehension, the most difficult texts are the writer-centred texts in, e.g., science and history, which often use the latest and even controversial research results. By this inclusion the writers ensure that the text represents their conception of the subject, but the interactive character of reading comprehension may be forgotten.

Current learning materials do not present the material in outline form, which would facilitate the reader's comprehension task. For example, textbooks do not contain metatexts concisely stating what will be discussed next. As the text proceeds, connections are not made between sections discussed earlier, nor are transitions to new themes clearly pointed out. Even in our textbooks of the lower level of the comprehensive school, narrative and descriptive material which would stimulate imagination is almost totally missing. It has now been decided to give more attention to this issue in Finland.

A third possible cause of the present situation is the home environment. Even today, mastery of reading comprehension in the Finnish comprehensive school is related to the mother's education and the number of books in the home. The more educated the mother is, the better her children's reading comprehension seems to be.

2.1.2.4 Conclusions related to the change of teaching

On the basis of Finnish investigations, three approaches have been identified which seem to produce readers who are capable of genuine interaction with the text. These approaches should be taken into consideration from the very beginning of reading instruction.

Firstly, reading comprehension and literature instruction should offer students different ways of approaching the text (personal, affective, analytical, interpretative and evaluative) and give them an opportunity to practise these in peace. It should also expose the student to the different text types and genres, encouraging individual responses to the same text, and pointing out that these responses may vary.

Secondly, literature reading and discussions of literature are vital to the development of reading comprehension, because they enhance imagination and experiences. One sign of improvement in reading comprehension is the fact that a student is willing and able to read more and more literature. Therefore, an effort should be made to provide more opportunities for reading in the course of instruction. Another goal which should be pursued at the same time is learning to use literature as the starting point for experiences and personal growth.

Thirdly, from the viewpoint of the development of reading comprehension skills, it is crucial to learn discussion techniques. More profound understanding of a text usually involves an ability and desire to discuss what one is reading. This ability and desire can only be developed if students are given time to express themselves either through general discussion or in group discussions, or by dramatizing or writing about the events of the text. When the text is thus transferred to the students' own expression they learn that the reading event is not over when the text is finished, but that it continues as a reflection and interpretation of the experience.

Another skill which is related to discussion is the ability to identify problems, which is also essential to the development of good writing skills. Good and bad readers and good and bad writers are often distinguished by this very ability. In order to develop this skill students must have experience in setting questions themselves. In teaching situations it is usual that too little time is left for students' questions.

The kind of instruction which has been described above regards discussion and problem-finding as the main objectives of teaching, and not merely as by-products of other objectives. In this type of instruction the teacher does not think for the students, and the students themselves participate personally in the shaping of knowledge and opinions. Also, the instructional materials needed for this kind of teaching will not have already chewed up the subject matter into seemingly small bits, but rather will offer different viewpoints on the same issue. The learning material should include more comprehensive texts, and students must learn to combine the facts contained in these texts.

PUBLICATIONS
(Only the publications focusing on reading have been listed)

Vahapassi A (1975) Centrally administered achievement tests in the academic year 1974-75. Tests and test results in mother tongue (Finnish) at the upper level of the comprehensive school. Reports from the Institute for Educational Research (IER), 58.

Vahapassi A (1977) The level of reading and writing in grade 6 of the comprhensive school, 1974-75. Reports from the IER, 88.

Vahapassi A (1977) On the structure and variability of reading skills in grade 3 of the comprehensive school in school year 1973-74. Reports from the IER.

Vahapassi A and Paunonen L (1978) Achievement in mother tongue in grade II of the High School during academic year 1973-74. Reports from the IER 117.

Vahapassi A (1982) The Structure of Reading Skills, The Annual Yearbook of the National Council of Teachers of Mother Tongue.

Vahapassi A (1982) How to educate real readers and writers, Annual Yearbook of the Association of Applied Linguistics in Finland.

Vahapassi A (1981) The first Situational Survey of Finnish Comprehensive School.
Reading Comprehension in ninth grade
Reports from the IER 188
Reading Comprehension in sixth grade
Reports from the IER 187
Reading Comprehension in third grade
Reports from the IER 186.

Vahapassi A and Takala S (in press) Reading Comprehension and Textbooks.

2.2 READING PROCESS AND LITERACY DEVELOPMENT

by

Birgita ALLARD and Bo SUNDBLAD
Institute of Education, Stockholm

Extract from the *Bulletin for Colleges of Teacher Education in Co-operation*

2.2.1 THE DEVELOPMENT OF READING ABILITY

2.2.1.1 General

We regard the development of reading ability as part and parcel of linguistic development. It can be illustrated as follows:

LANGUAGE DEVELOPMENT

Individual	_____	"Motherese"	_____	Surroundings
Individual	_____	Speech	_____	Surroundings
Individual	_____	Reading	_____	Surroundings

All three forms of language develop through the individual's interaction with his or her surroundings. This development can be delayed or given optimum opportunities, depending on the quality of this interaction. Language development, of course, is closely bound up with the child's overall development. Feelings, thoughts, needs, experiences and action in everyday situations are both the precondition and the content of communication. The development of every form of language is a process over time. From an adult point of view, the child does not develop a decent use of the spoken language until it is six or seven years old. But this development continues for the whole of the life cycle. Reading development is exactly like speech development in that it begins long before the child can directly apply it. And from the moment the child begins applying its capacity, i.e. starts to read on its own, this development will, at best, continue for the rest of the child's life.

In order to arrive at a better understanding of the reading development process, it is important to know about the reading process itself, i.e. what we do when we read. If you ask the first person you meet "What do you do when you read?" the answer will probably be: "I read the letters and put them together into words. Then I recognise the words and know what they mean".

This can be illustrated as follows:

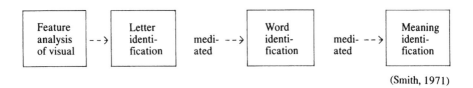

(Smith, 1971)

Reading research has long assumed that this is how reading proceeds in all situations. This is a superficial, mechanical approach which excludes the operations of understanding and processing what is read.

We can advance one step further by completing Smith's model:

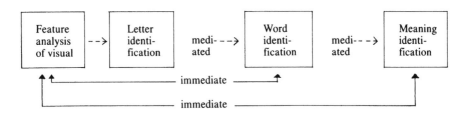

Mediated reading occurs in connection with particularly difficult and incomprehensible texts, but the usual procedure in connection with the continuous reading of intelligible text is indicated by the arrow from the first to the fourth panels.

2.2.2 A MODEL FOR THE READING PROCESS

Let us go one step further and present the following model, which describes the factors included in the reading process (Edfeldt, 1982, Sundblad et al 1981).

126

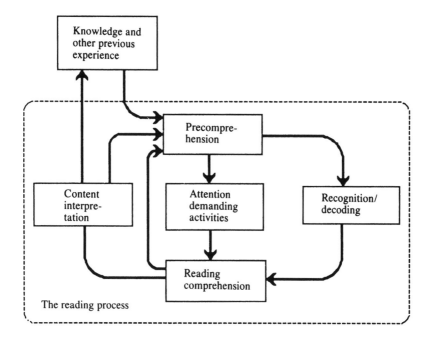

The reading process

2.2.2.1 Previous experience and precomprehension

Whenever we are confronted with a written passage, we "know" something about it: somebody gives it to us, we see a picture, we see the title and so on. We do not need much in order to have some idea of what the text contains.

Mere knowledge of the supplier, for example, is sufficient to point our thoughts in a certain direction. This is because we always utilise our previous experience and knowledge, especially when confronted by a reading passage. This previous knowledge naturally also includes our linguistic ability. Marton *et al* (1977) summarise this accumulated fund of knowledge and experience in the term "world image". The world "image" suggests that we are not dealing solely with a collection of facts but also with the individual person's way of handling and processing impressions and lessons and the result of these operations in the form of emotions, expectations and ideas. We have all had experience of attaching emotions to certain kinds of reading matter. We can experience a tremendous reluctance to read a passage which we believe is too difficult and incomprehensible or which we feel forced to read, while on the other hand we can rip through very difficult, complicated passages dealing with a topic which captures our interest. Both the reading situation and the actual business of reading are coloured by our world image.

We thus have a permanent preparedness, in the form of extensive previous knowledge, for receiving a message. This applies both to children and to adults. For present purposes we will confine our attention to adults with a pre-established reading ability.

The instant reading passage is put in our hand, whatever it may be (a book, a newspaper, a letter, a note, or the subtitles on the television screen), our previous knowledge is activated in relation to it. We are prepared, for example, as soon as we unfold a newspaper. In a sense, one can say that we are already familiar with the newspaper. We have a general, approximate idea what is usually on the front page. If, in addition, we have been following the progress of events, we can quite safely predict what news items will be included.

Our expectations regarding the content of the daily paper govern the activation of our previous knowledge, i.e. they determine what may be relevant to retrieve for present purposes. We can check this by the back door, so to speak, not least with reference to situations where we have retrieved the "wrong" things and they have clashed with the content of the reading passage, i.e. when we have been expecting something different from what we actually read.

This activation or structuring of certain items from our previous knowledge and experience we call *precomprehension.* This brings us on to the actual reading process, because a reading passage always builds up preparedness in the form of precomprehension. In the above diagram, "previous experience" has been placed outside the reading process while precomprehension has been included in it.

2.2.2.2 Precomprehension and recognition/decoding

This brings us to the actual process of decoding — that part of the reading process which many reading researchers and educationalists have regarded and still regard as identical with reading, i.e. they confine the reading process to mere decoding. Recognition or decoding is the registration by the eye of the letters, words and phrases in the text.

Our nervous system includes an efficient mechanism for determining priorities. That part of the brain called the RAS (reticular activation system) is vitally important. When we see or scan something, this operation has already been preceded by a series of unconscious capacity for filling in the spaces between the characteristics we use. For example, you know it is Judy when you see the red beret, the green jacket and the lilting gait, you recognise her, "all of her", although you have perhaps only seen three characteristics.

The same phenomenon applies to reading identification, and we ourselves are convinced that we have read every single letter and word, even though we have not needed to do so. Our brain has an ingenious way of working. First extremely rationally, with a minimum of characteristics, and filling in what is needed in order for us to grasp the whole; this is made possible by the function of precomprehension. We are not aware of this happening, and so we believe that the actual course of events is quite different. In a manner of speaking, our brain is constantly deceiving us: we do not need to read every single letter, even though we think we do. This is yet another example of the immensely creative nature of our thinking.

Vagueness and ignorance of the way in which our brain thus operates naturally lead to a misconception of the role of decoding in the reading process, so that decoding, and this alone, is taken to be reading, and is thus divorced from its natural and obvious context. This misconception causes the practical development of a decoding technique to be given a completely mistaken focus.

The apparently straightforward word "decoding" thus conceals a very complex process. Referring back to the phenomenon of precomprehension, we can now see that this decides what we will scan or "see". It is on the basis of precomprehension that we select characteristics relevant to the context. Thus the interesting thing is that our previous knowledge influences the actual decoding process via precomprehension. In other words, in order for the recognition of characters in a continuous reading passage to be capable of forming a message, we have to have a preparedness in the form of an assumption concerning the content of the message immediately following.

2.2.2.3 Precomprehension and attention-demanding activities

Just now we referred to the ability of the brain to "fill in the blanks". The way in which this is done is very much controlled by precomprehension.

The notional content of precomprehension is often quite rough and ready. Further processing and structuring then takes place. This process also has the effect of making the notional content accessible to our consciousness if needed.

One can speak of two parallel processes, and we shall try to elucidate them by means of an analogy. Electrical impulses passing through the air or a cable are decoded and become an image on a television screen. Recognition is a comparable process of decoding, supplying the brain with an image. But in order for the image/decoding to be understood, another analytical process is required — a further development of the content and structure of precomprehension. The habitual reader need not be aware of this analytical process as long as the reading process is unimpeded.

This part of the reading process is extremely many-sided and it is therefore difficult to find an adequate name for it. The term "attention-demanding activities" stresses the conscious thinking involved by the reading process, but of course most of the work is done unconsciously.

2.2.2.4 Reading comprehension

Reading comprehension requires both form and content — both processing of precomprehension and decoding. Without the processing of precomprehension we are left with mere "empty" words, and without decoding we are left with mere conjecture.

One of these two routes to reading comprehension, that going by way of decoding, is fully automatic in the habitual reader, i.e. linguistic reception via characters is, so to speak, in the reader's blood. Comprehension analysis, however, can never proceed automatically, though it can proceed unconsciously as in the case of the habitual reader.

If the reader does not need to concentrate on reading comprehension, which after all is only a small unit of understanding, it will remain unconscious.

Summing up, reading comprehension is the combined result of pre-comprehension, analysis and decoding, and can be either conscious or unconscious. In the case of the habitual reader, it proceeds quite unconsciously and automatically.

2.2.2.5 Reading comprehension, content interpretation and precomprehension

Reading comprehension is limited in scope. It can range from parts of a word to a short paragraph. These units of reading comprehension provide material with which to develop continuous precomprehension and also content interpretation. Content interpretation is not built up piece by piece. This is already apparent from what we have said concerning the ingenious working technique of the brain. We are able all the time to construct wholes from quite a small amount of information material. What happens when we develop content inter- pretation is that we acquire a better and better picture of the whole. The holistic image constituted by content interpretation provides the unconscious precomprehension with an uninterrupted flow of information all the time reading is in progress.

The habitual reader can choose whether or not to concentrate on content interpretation.

2.2.2.6 Content interpretation in relation to knowledge and other experiences

Thus the reading process includes the development of a content interpretation which is essential to continued reading. But content interpretation can also be a result of the reading process. No matter how we have read, content interpretation is something we can retrieve from the "ongoing" reading. At any time during the course of reading, we can relate this overall appreciation of content to our other knowledge and accumulated experience in a conscious process of thought. When we think about content, we probably always do so in relation to what we already know and are capable of. The processing of content can be a highly creative process.

The content of content interpretation leaves the area delineating the actual reading process, thus closing the circle.

130

2.2.2.7 Focussing attention

The whole of the process we have described so far, illustrated by means of the figure above, is unconscious and partly automatic (decoding) in the case of the habitual reader. But when we read we are still conscious of reading and conscious of what we are reading. Our attention is not focussed on any part of the reading process, otherwise we would not be able to get anything out of our reading. This focussing of attention is based on the contentive processing of precomprehension and on the purpose of reading.

The habitual reader can focus his attention on the relationship between content interpretation and previous knowledge, especially when reading non-fiction. In some cases his attention can focus on actual content interpretation, for example when reading a good novel. Earlier we stated our belief that reading in most cases involves focussing attention on content interpretation in relation to what we already know and are capable of. This also applies to the reading of novels. The difference in the focussing of attention in the two novel-reading situations lies in our knowledge and experience of the genre, the author and so on. If we are familiar with the type of novel concerned, we concentrate our attention on its actual content. But if the novel is unfamiliar in several respects, our attention will probably focus on the relationship between content and previous knowledge.

Attention can also be made to focus on reading comprehension in situations where we have difficulty in understanding a message. It can also be made to focus on scanning, as in connection with difficult words or abstruse text.

We could illustrate the focussing of attention by imagining a moving needle or pointer which pivots in the "attention-demanding activities" panel and can be trained on any of the other parts of the reading process.

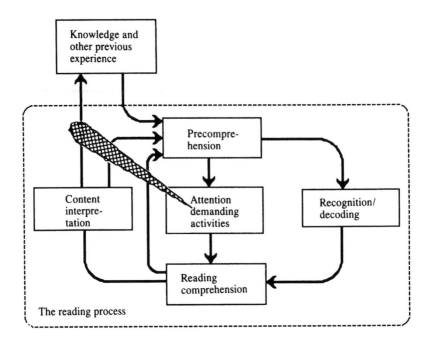

The reading process

2.2.3 CONSEQUENCES FOR LITERACY DEVELOPMENT

2.2.3.1 General

Considering these two views of the nature of reading as phenomena, i.e. reading as mere decoding and the model for the reading process, we find that they lead to completely different approaches to reading instruction.

Starting by looking at letters, identifying them and after putting them together identifying words and so on, describes a synthetic procedure which implies that initial reading instruction should start with the learning and identification of letter and sounds, followed by learning how to combine them in "simple" short words. This learning acquires the character of drill or, putting it differently, decoding has to be automated with the aid of overlearning. The comprehension aspect is not a primary but a secondary benefit so far as pupils succeed in understanding the so called short, simple words — which are not always even included in their passive vocabularies. It is not until the pupils have mastered decoding that they reach the point of reading passages for the sake of their content. This description fits the traditional and commonest method of reading instruction.

From the very beginning, teaching to read concentrates on the meaning of the *symbols* — not on the message and meaning contained in the text. During his earliest years in school the child receives the impression that skills such as decoding are what reading is all about. Some children will achieve literacy despite the experiences to which they have been exposed in school, illiterate persons unable to find reading truly meaningful. In a technical sense they can read — they master the technique of decoding.

It can be illustrated as follows:

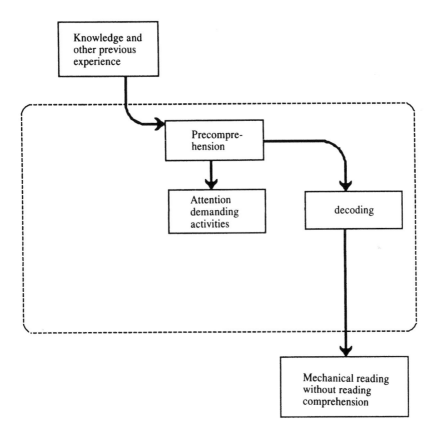

The reading process model has completely different consequences. Applied to the situation of speech and listening, it will be as follows:

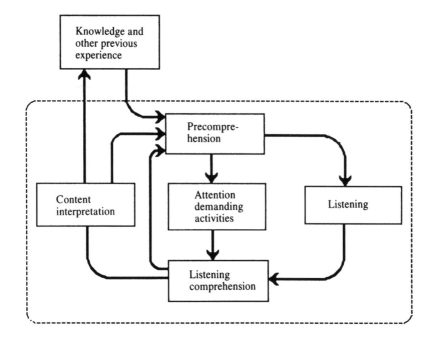

Thus the process is the same. Reading development rests on and emanates from the speech process. The knowledge of a child about to learn to read includes a linguistic mechanism which, if not utilised in the encounter with the written language, will impede or indeed sometimes preclude the mastery of reading.

The principle is therefore:

Familiar, continuous passages — which can very well be composed by the pupil himself — closely resembling the form of the spoken language. In this way decoding will not pre-empt all attention but will be facilitated by precomprehension being given the opportunity of acting. During the initial period the pupil always reads aloud. This means that the linguistic mechanism for speaking and listening sustains and facilitates decoding by communicating context and understanding.

On the basis of this knowledge concerning reading, we have collaborated with teachers in drawing up a reading development schedule which comprises 23 points and proceeds from the child's initial interest in letters of the alphabet, to the most advanced reading obtainable by an adult. The schedule is divided into three phases: "the discovery phase", "the consolidation of reading experience" and "active text processing". (Allard and Sunblad 1982). The final phase could also be termed "development of good study capacity". The schedule focusses on what the pupils are capable of, not on what he or she cannot do, and it is a qualitative implement.

The aim is to supply teachers with an aid for discovering and taking an interest in the character of the reading process. But the schedule has also proved an excellent aid to a diagnostic approach. Its greatest merit is that it bridges the gap between theory and practice.

REFERENCES

Allard B and Sundblad B (1982) *Handbok om lasning:* Stockholm: Liber Utbildningsforlaget.
Edfeldt A W (1982) *Lasprocessen — grundbok om lasforskning:* Stockholm: Liber Utbildningsforlaget.
Marton *et al* (1977) *Inlarning och omvarldsuppfattning:* Stockholm: AWE/Gebers.
Smith F (1971) *Understanding Reading,* New York.
Sundblad B, Dominkovic K and Allard B (1981) *LUS — en bok om lasutveckling:* Stockholm: Liber Utbildningsforlaget.
Extract from Sundblad B *et al* (1983) "LUS — en bok om lasutveckling" Liber Utbildningsforlaget, Stockholm.

THE STRUCTURE OF LUS

The Literacy Development Scheme (LUS) comprises 23 points indicating the development of reading ability from the very first interest in letters to maximum development. Each point represents an identifiable step in an ongoing literacy-development process.

The 23 points of the schedule are divided into three sections, interspersed with text passages.

Points 1-10 describe stages of basic literacy development.

Points 11-19 show literacy development on the basis of reading which is still partly phonetic (point 11), proceeding to the stage where the reader thinks more of what he is reading than of the mere fact of reading (point 18).

Points 20-23 describe how the reader acquires such a command of reading as to be capable of reading any book whatsoever with a result corresponding to this actual need.

Several points in the LUS are preceded by an italicised text in an oval or oblong frame.

The ovals indicate the insight of knowledge which the child has developed. The child itself cannot formulate this insight, because the insight is unconscious.

But one can conclude from the child's way of dealing with a reading passage that a certain insight has been achieved and constitutes the very foundation of the reading ability shown in a stage of literacy development.

The oblong panels show some of the prerequisites of literacy development. These are concerned with the will to read and the pupil's experience of himself in relation to his reading. The desire to read is important at all stages of the child's reading development, but we believe it to be more important at some stages than at others.

HERE, THEN, IS THE LITERACY DEVELOPMENT SCHEDULE (LUS)

LITERACY DEVELOPMENT CHART

1   Finds ("reads") and draws ("writes") his name.

2   Knows the direction of reading, i.e. that one reads from left to right, from top to bottom, and how to change lines.

> Realisation that writing is
> recorded speech

3   Knows that a written letter has a sound.

> Realisation that one can distinguish between
> the content of a word and the word as such

4   Can split up and put together a word which is important to him.

> Realisation that one can obtain a meaning,
> that the result can be a word,
> when several sounds are put together

5   Can run short or familiar words together

> Realisation that one can read complete words
> without identifying sounds one by one

6   Can read ordinary short words in reading passages, e.g. and, a, an... directly with the aid of contextual understanding.

> Realisation that one is always reached
> by a messenger via reading

> Ability to concentrate on a reading
> passage for a few minutes

7   Can get through a simple reading passage with what for him (i.e. the pupil) is an appropriate content apprehension. Understands a simple written instruction.

> Realisation that one can communicate
> by means of written text

8   Can substitute communication by means of the written word (reading-writing) for a listening-and-speaking situation.

> Realisation that one letter can
> have more than one sound

9   Can read words with double letters and some deceptively spelt words by means of content apprehension.

137

10 Can also put together longer and unfamiliar words.

*

> The will to learn – children's
> own motive for reading learning

11 Sounds but reads more and more words without sounding. Uses reading comprehension.

> Wants to read more advanced text and is
> in a hurry to get on

12 "Fast sounds" but gets stuck and sounds certain relatively simple words.

> Given support, self-confidence leads to necessary practice

13 Reading more fluent but still quite a few mistakes.

14 Can understand a working description (extensive written instructions), e.g. a recipe or other instructions in several stages.

15 Reads fluently.

> Realisation that one can look for a certain
> word (or content) and detach oneself from
> the rest of the passage

16 Can scan-read, i.e. rapidly find individual items of information.

17 Can understand the content of a foreign film on television, using the Swedish subtitles.

18 Reads a lot. Sticks mostly to one kind of book. Reading is governed by precomprehension and no attention has to be paid to the actual scanning process.

19 Can read extensively, i.e. size up the main content of passage by rapidly skimming through it.

*

```
┌─────────────────────────────────────────────────────┐
│         Reading of different types of reading matter, │
│          e.g. different kinds of fiction, non-fiction,│
│                      newspapers                       │
└─────────────────────────────────────────────────────┘

┌─────────────────────────────────────────────────────┐
│            Ability to steer and control one's own     │
│                      thinking                         │
└─────────────────────────────────────────────────────┘
```

20 In-depth reading. The reader can deliberately enter into close and detailed contact with the content of the text in order to get as much out of it as possible.

21 Good reading of various kinds of fiction and simple non-fiction.

22 Can consciously adapt the mode of reading to the character of the task, the purpose in hand, the text its content, either spontaneously or when called upon to do so.

23 Unconsciously adapts the mode of reading to the character of the passage and to the total reading situation.

2.3 WRITING – A WAY OF THINKING

by

Birgita ALLARD and Bo SUNDBLAD
Institute of Education, Stockholm

2.3.1 SUMMARY

This article consists of some excerpts from one of our four books. The common
theme of all four is "Writing — a way of thinking" and the text below is taken
from book 1, chapter 2, entitled "When we read and write...", which discusses the
writing process. In it, we make a distinction between the creation of the written
message and the revising of the written message. Here we are only going to take
up certain aspects referring to the creation of written language, based on our
model of the writing process (figure 1).

2.3.2 THE MODEL OF THE WRITING PROCESS

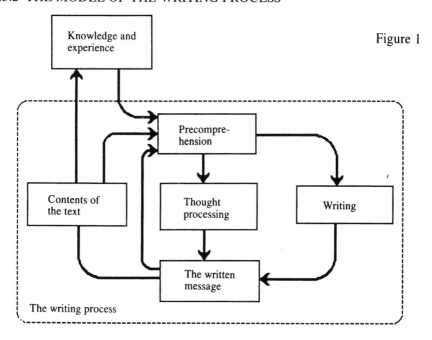

Figure 1

2.3.2.1 Knowledge and Experience

What we have said about "knowledge and experience" in connection with reading also applies here, of course. The dotted "frame" above, summarizes everything that makes up a human being. It goes without saying that his experience of writing, his conceptions of written language as a form and as a means of expression and his perception of his own competence as a writer are included within this frame. The majority of people associate learning to write with school. One's experiences from one's school days colour and even determine each individual's ideas of what writing is all about, what it is for and how it can be learned. An example: three girls in the 6th form were asked what it is to learn to write. They answered by listing the following: "the ng-sound; the sj-sound/ spelling/: when words are spelt with a and not e; understanding difficult words; when you say him and not he; verbs and nouns and such like"; and so on. When we asked "What about essay-writing then?" they answered "Oh, you mean free writing, which we do for one hour on Fridays! That's so boring!". What the girls are doing is enumerating the different subjects and parts of their text-books in the mother tongue (grammar). In addition to the subjects they mentioned, like spelling rules, comprehension, parts of speech, etc, they demonstrate the invisible side of learning, namely that writing is defined according to these subjects, and equally that one's own competence as a writer is defined according to how well one achieves in them. Someone who is good at writing is, for example, someone who always spells correctly, and vice versa.

A good friend told us that she had begun an evening class in Swedish — the mother tongue. One evening they had to write an essay. Karin said, "Oh, what fun it was! I wrote and wrote, page after page and gave it in feeling happy and pleased with myself. The following week the teacher came up to me and said that her whole weekend had been hellish because of me. She had slaved her head off trying to get my essay into correct Swedish. Though the contents were good, she said. What I had written could hardly be seen through all red correction marks. That was how my efforts at writing turned out. But then, I have never been able to write, what with being "word-blind". But that was the first time I had ever thought it was fun." "What did you do with the essay?", we asked. "Threw it into the waste paper basket, of course!".

Our experience of the written language — both through reading and writing — influences us, even though it may be unconscious, in all writing situations. To write means that one is the source and author of the contents of a text and it is from one's own knowledge and experience that one draws the material.

2.3.2.2 Pre-comprehension

As soon as one gets an idea — a thought that I want, or ought to write about this — the actual writing process has begun. We are then at the stage of pre-comprehension. The idea, subject or heading sets in motion this unconscious process, in which certain aspects with regard to content, language and form are brought to the fore. A particular aspect may appear as a conscious thought or idea, and becomes the thread which one can begin to unravel, whilst the rest remains in the unconscious. Irrespective of how the writing process begins, either by imposition from the outside or on one's own initiative, a mobilization of this pre-comprehension occurs, the cources of which are the sum total of the individual's knowledge and world of experience. This means that the variations between individuals are countless. When a mass of people are given a subject or a title to write about, no single one of all these written products will be identical with another either vis a vis their content or their outward linguistic appearance. If the subject is a factual one, there may be some factual points of contact between them, but beyond that they will not be alike.

Pre-comprehension prepares the way for the writing process by giving it a general direction in which one's aims, the content of the text, linguistic aspects, and ideas about those reading the text are interwoven with each other. As the author, one has maximum pre-comprehension, one is "inside oneself", even if this is still largely unconscious and visible only as an idea, an image, or in the form of a word that summarizes the totality. It is something of a paradox that almost all of it is unconscious and yet one knows it all. Thanks to the conscious idea and the unconscious process that this idea sets in motion, one knows, perhaps more in the form of intuition that anything else, roughly what one is going to write about. Despite this, one cannot predict what the finished text will contain.

2.3.2.3 Thinking – (speaking) – Writing

Letting the idea (pre-comprehension) take shape — by writing — is a process that is insolubly linked with thought, feeling, and imagination. The creation of a written message is more a question of thought than a handicraft. Even such a simple situation as writing a shopping list requires one's thought. One may perhaps have a dialogue with oneself: What do I need to buy today? What is missing? What are we going to eat for dinner? As soon as one thinks something, one writes down. The general idea "we must do some shopping today" is formed on paper in detail through the thought process. But how one thinks depends on one's knowledge and experience — how good a cook one is, one's likes and dislikes, one may even imagine the taste as one is selecting items for the list.

Everything that you are as a person governs the thought processes that become visible in what you write. Without consciously wondering about it, your shopping list takes the form that shopping lists usually have — you have made a real list. But the order in which you have written down the items is completely personal. One person may envisage the shop in front of him and writes down the different items according to where one can find them in the shop. Someone else may write according to the ingredients in a recipe. Each person has his own way of organizing and structuring his thought. Thus the written message is a product of thought processing and writing, as the model shows.

Before proceeding further, we shall make it clear what we mean by the words "thinking" and "to think". Normally one understands thinking to mean something rational, non-emotional and free of irrational motives and needs. We, however, start from the assumption that human functioning does not allow itself to be split up into mutually exclusive compartments, but that all these aspects function simultaneously. Of course, we can consciously focus on, study and choose to describe a particular aspect, but we always function as totalities, irrespective of whether we are aware of it or not. This means that our thought processes comprise all the dimensions of our lives. Both our past history and our present lives are reflected in them. Everything that we have ever learned and are still learning, everything that we have experienced in the past and experience now, is always connected with situations and people. The thought processes are therefore always emotionally charged to different degrees and are the basis of our opinions and values, of our conception of ourselves and of other people, of our morals and ethics, and of our motives and actions. Even if our thoughts are related to our inner selves, they include the world around us as we have confronted it in the past and continue to confront it. Everything that via thought is expressed linguistically (as inner, spoken or written language) makes visible for us who we are, what we can, want, think, hope, believe, and so on. One usually understands knowledge as being something neutral - facts are facts, and nothing else. But knowledge can never be isolated from the learning situation as a whole and the way in which the individual interprets and experiences it. Knowledge always has emotional undertones. (See further book 2, chapter 1).

Let us return to writing. What one writes, and for whom one writes, varies — diaries, letters, memoranda, debate articles, debate books, factual articles, non-fiction books, novels, poetry, and so on. Certain writing situations are more demanding than others. There is, for example, a considerable difference between writing a letter to a good friend and appealing to the tax authorities against a refused tax reduction. The difference lies both in aspects of form and content. Coupled with different writing tasks are different generally accepted require-ments as to form, or conventions about the length, linguistic style, arrangement and so on. An experienced writer adapts himself instinctively to the given situation. Yet despite these outer limits, there exists a certain freedom as to content and language. (We return to this in book 3, chapter 3).

The creation of the written message and the way in which thoughts become words and sentences on paper is an almost totally neglected aspect in connection with learning to write. We start out from the premise that the same factors (see the model) function, irrespective of whether one is in the process of learning to write or already can write. The differences can be found in the contents of the frame "knowledge and experience".

2.3.2.4 *The relation between inner and outer language*

The relation between thought and language has been discussed in different ways from different theoretical points of view. We have chosen to base the discussion on Vygotsky (1969, 1982), since we have found his position to be very fruitful. One of his points of departure is that thought can be both non-linguistic and linguistic. What he discusses primarily is linguistic thought which he also calls "inner speech". The link between thought (inner speech) and language (speech and writing) is found in the meaning of words, or as he expresses it:

> "Word meaning is a phenomenon of thought only in so far as thought is embodied in speech, and of speech only in so far as speech is connected with thought and illumined by it. It is a phenomenon of verbal thought, or meaningful speech — a union of word and thought." (1969, p.120).

He compares inner speech, speech and writing with respect to the way in which the meanings of words are expressed in sentences for the respective forms of language — thus he compares their syntax (sentence structure).

Let us begin with the following colloquial situation. Five people are sitting round a table drinking coffee. Eva looks at Erik, reaches her hand towards the coffee pot and says: "More?". "No, thanks", replies Erik. This single word "more" expresses a whole context of meaning which everyone present understands. The context of the situation, Eva's gesture, and her questioning tone of voice give meaning to the word. Expressed in our terms, we can say that everyone sitting round the table has a mutual pre-comprehension given by the shared situation that they are taking part in.

Sometimes, however, the outer situation is not so self-evident and clear, and yet people who know one another well can understand each other with a minimum of words. What characterizes the spoken language is the possibility to abbreviate the sentence — to reduce the syntax — which is possible precisely because the meaning of the words can be conveyed by a number of non-verbal factors such as gestures, looks, tone of voice, context and so on.

> "A simplified syntax, condensation, and a greatly reduced number of words characterize the tendency to predication which appears in external speech when partners know what is going on." (ibid, p.141).

This does not apply to all spoken language situations. Sometimes we have to be very detailed, for example when the person we are talking to does not understand the situation or the subject, or when one talks to an unknown person on the telephone.

When we write, we do not have a conversation partner with us. We cannot take for granted that the reader of our text has prior knowledge of the theme or the context referred to in what we write. Things, which in a conversation may be expressed by non-verbal means, must in writing be expressed in words and sentences. Shortening the sentence is not possible except when we scribble down something meant only for our own eyes. The written language is the language form that uses the most words and that has the most precise and developed syntax. Thus, if we compare the spoken with the written language, we find that the tendency to abbreviate sentences that is possible in many colloquial situations is out of the question in written language. If we then turn to inner speech, we find that this tendency is the rule. Or as Vygotsky (1981) says:

"This tendency/toward abbreviation and predication/never found in written speech and only sometimes in oral speech, arises in inner speech always./.../We know what we are thinking about — i.e., we always know the subject and the situation. Psychological contact between partners in a conversation may establish a mutual perception leading to the understanding of abbreviated speech. In inner speech, the "mutual" perception is always there, in absolute form; therefore, a practically wordless "communication" of even the most complicated thoughts is the rule." (p.145).

The syntax of outer language (spoken or written) is dissolved in inner speech and we find in inner speech a form that is completely distinct from the outer. In inner speech the meaning of words comes to the fore.

Within the framework of a certain common language, a word has its specific and relatively stable meaning. Shifts and changes of meaning are quite a lengthy process. The meanings of words can therefore be set down in dictionaries and lexicons. But as soon as we use a word in an entirely linguistic connection, it can take on a far richer significance than its mere dictionary meaning. In, for example, myths, fables, fairy tales and parables, the literal meaning acquires a symbolic significance. In our inner speech, the dictionary meaning of the word is completely subordinate to the "meaning" we ourselves give to it. We charge the word with personal meanings, according to our experiences, emotions, ideas, interpretations, and the context in which we exist. The word for us has both a far richer significance in terms of the manifold situations to which it is related, and a narrower one, in the sense that our experiences are personal and subjective. This significance or implication alludes to whole complex situations where the word itself functions as a hint. Vygotsky (1981) draws the conclusion:

"Thought, unlike speech, does not consist of separate units. When I wish to communicate the thought that today I saw a barefoot boy in a blue shirt running down the street, I do not see every item separately: the boy, the shirt, its blue colour, his running, the absence of shoes. I conceive of all this in one thought, but I put it into separate words. A speaker often takes several minutes to disclose one thought. In his mind the whole thought is present at once, but in speech it has to be developed successively. A thought may be compared to a cloud shedding a shower of words. Precisely because thought does not have its automatic counterpart in words, the transition from thought to word leads through meaning." (p.150).

When we try to express a thought in words, we can never do it exactly. The same thought can be expressed in different sentences, just as a sentence already formulated can be the expression of different thoughts.

"The structure of speech does not simply mirror the structure of thought; that is why words cannot be put on by thought like a ready-made garment. Thought undergoes many changes as it turns into speech. It does not merely find expression in speech; it finds its reality and form." (p.126).

146

This means that the path from thought to words, and especially to written words, is long and winding. It is necessary to capture these momentary, complex thought structures and stay inside them long enough for them to become words on paper. Sometimes this may seem impossible; there are not enough words or there does not seem to exist a word for what we want to express. The words and sentences that we use are never completely adequate expressions of our thoughts, and they never say everything. There is always something unexpressed, something between the lines.

2.3.2.5 Language makes thought visible

A thought as a complete complex structure can be illustrated with the following example. Let us say that a memory suddenly comes to mind, clearly and sharply. You can see the whole situation in front of you, in all its details: the room, furniture, ornaments, the people, their clothes, faces and expressions, the vibrations, undertones, relations, and perhaps even the smells and your own feelings and mood in the situation. All this in less than a second. A thought can thus be an image which includes everything. Irrespective of whether we think in pictures or not, our thoughts are always totalities, more or less clear and more or less detailed. They can flash by and disappear, frustrate or irritate us. It is very seldom that they appear with complete clarity.

Thoughts or momentary images are not always coherent and we are not always conscious of all their aspects. In order to give order and structure to what flickers by, we may try to "talk" within us. We formulate a sentence or the beginning of a sentence, perhaps only the hint of a sentence. Before it is formed, something else crops up. "What was it now again? Where was I?" Perhaps you recognize the following situation. You are filled with problems which spin round in your head. You feel confused, it is hard to come to grips with it, and it seems impossible to solve. The same details crop up time and time again and you cannot get a grasp of the complete picture. A good friend comes to visit and you wonder if you can have a word with him. Then you start to formulate the problem and this vague, compact, impossible "thing" starts to take shape in the form of words — it becomes visible to you within the framework of the linguistic structure. Your friend occasionally interposes: "Ah ha! Oh yes! How?" Suddenly the picture becomes clear — yes, of course, that's what I shall do! And you thank your friend for his help. You believe that it is thanks to your friend you have solved your problem. You do not notice that you helped yourself by formulating and by bringing out into the open the whole extent of the problem. But of course, it is easier to talk when there is someone there to listen. You could actually have managed by writing it down.

Thoughts are born through words — more so when we write because they are always there to go back to. Thoughts become visible and we get to know what we are thinking. Language is like a map which discloses and gives us perspective over the landscape of our thoughts. It is absurd to think that everything that we have experienced and learned over the years should be borne with us on a conscious level all the time. Fortunately, we do not function like that. Almost everything is under the surface, but a good deal is comparatively easy to bring to the light in conscious thought. But it only becomes fully visible once it is clothed in words. Actually, we do not always know what we know and understand about many things. The fact that our knowledge is largely invisible and hidden, so long as we are not thinking about it or formulating it, makes us sometimes think that we do not know anything at all about certain phenomena or that we have only vague ideas. Instead of saying "I can't", or "I don't know", it would be more adequate to say "I don't know what, or how much I know about it, or how thoroughly I know it", which implies "not until I have had time to think about it".

Let us imagine that some persons have been given the task of writing about "Roses" — everything that they know and have learned about them. When all the possible opposition and objections, such as "I can't!" and so on, have been overcome and they have started and been given a little time, it is highly likely that some of them will surprise themselves. Astonished, they will discover that they have produced more than they thought was possible and that they can, if given more time, find even more to write about the subject. At first, one thinks that there is not anything to write about on this subject, or at any rate very little. Then, after one has started to write, more and more begins to crop up. It is this that makes writing so fascinating: to know the rough outlines in advance, but never to be able to anticipate the results; that a sentence written down suddenly leads one into a new and unexpected train of thought that one did not know existed. Out of which one experiences as nothing, something is created and formed. Maybe it is this, among other things, that Ivar Lo-Johansson expresses (1981):

> "I suddenly felt that I should now write down what I had been bearing within me for so many years. I had carried its memories along with me like a kind of invisible luggage. It had weighed me down, although for other people it had been invisible. Others had only seen that I had sometimes been brooding over something. I had also felt happy with my burden. — What are you thinking about? someone occasionally asked. — Nothing especially, I had replied. To a certain extent, this was a truthful reply. It was what I had not been able to account for that I would now describe. As far as poetry is concerned, there existed nothing until it had taken its black shape on the white paper. Poetry is born from an apparent nothingness." (p.15).

2.3.2.6 The thought process

If we revert to the model, we can thus see that the coming into being of thought by the words on basis of a vague idea — pre-comprehension — results in a written message. After only a few words or sentences we can discern a direction, a content of the text which builds on and gives pre-comprehension more substance, and which also tells us something — we learn something about ourselves. This is the writing process in all its aspects.

This process can come to a halt or be interrupted, as well as continue to run smoothly. The latter aspect has been described by some authors as though the characters start to live their own lives and one needs only to go along with them. But sometimes the flow suddenly ceases. Very likely everyone knows how that feels. It feels as though one's head is empty. One writes two or three words, but they do not say anything — they are empty. Panic threatens. Those who are fairly experienced at writing know that this is not very serious. Perhaps they leave their writing to do something completely different. Occupied with something else, or after a night's sleep, the theme suddenly emerges again and the writing process can be resumed. Even if one leaves the writing desk, one does not leave the thought processes — they continue on unconscious levels. How one can cooperate with these unconscious processes is described by Lo-Johansson (1981) thus:

"I have a trick which I can perhaps teach others. In the evening before I go to sleep, I usually draw up a little plan of what I shall write the following day. It is just as if there was a meaning with it. The unconscious mind seems to busy itself with the material during the night. I wake up and think that a number of problems are solved. Sometimes I dream about the matter while I sleep. This means in fact that a writer works even when he is asleep." (p.132).

The creation of the written message thus has an invisible side that we can never completely understand or wholly grasp. We can never describe in detail what happens, nor how it happens, when an idea begins to work unconsciously. We can perceive its effects — that something has happened. Nevertheless, this aspect must neither be mystified nor ignored.

This invisible side of writing is, for the most part, neglected. The product itself counts, not the way in which it took form. When writers describe this aspect we take it to mean that it only applies to professional writers, not to ordinary people who only write occasionally. When we get irritated about our inspiration drying up on us, when our thoughts play tricks on us, we blame our frustration on a kind of general inability to write. We usually tend to regard things that we are not aware of as non-existent. If we have never been shown by anybody and we have never experienced it ourselves, we have difficulty in giving ourselves the necessary space for our thoughts. Nor do we dare to trust our unconscious processes — i.e. that the theme or the idea sets an activity in motion in spite of the fact that we may be busy with something else. Instead we react with impatience; we feel impotent and powerless. We may force ourselves to write something that we are not at all satisfied with. But it is absurd to believe that professional writers, and those who write occasionally, should be different concerning the nature of the thought processes, conscious and unconscious. The differences lie solely in the extent of writing-experience and in the fact that the professional writer both recognizes and depends on these processes.

Here, perhaps, we may be misunderstood as meaning that all one has to do is to hit on an idea, then leave it alone to work for us and wait for inspiration. We certainly do not mean this. It is the conscious effort of thinking-writing that opens up the unconscious level and sets it to work. At the same time this hidden and latent process is what gives life and substance to the contents of a text. Expressed in another way, one can say that these conscious and unconscious processes mutually presuppose each other. The result — what goes down on paper — is not always brought about through conscious thinking. Unconscious aspects may have a "direct line" to the pen or typewriter. When you read through what you have written, you may discover a train of thought of which you were totally unaware whilst you were writing: Did I really think and write this? That the writer may be surprised by what has gone down on paper is thus not surprising. Writing is never a one-sided intellectual or solely conscious activity — even if in certain cases we prefer to think so. Let us quote some more lines from Lo-Johansson (1981):

> "Whilst I had been writing I had suffered with the story of my novel. The anguish in what I had written had sometimes been so strong that, even for me, it came near to the unbearable. Sometimes I had to rush up and open the window in order to get some air, while I was struggling through certain parts of the book. When I was typing it out I had to do my very utmost to force myself to hit the keys on the typewriter." (p.106).

To what extent our writing is emotionally affected depends, of course, on the subject. Even an intellectual subject can awaken unexpected emotional reactions. But again — this may seem so absurd and unlikely that we hardly connect those reactions with the subject of our writing. Instead we put it down to tiredness, overwork, or a cold coming on. Perhaps we do not notice any physical or psychological reaction at all. And yet certain emotional charges and undertones are quite apparent in the text, as to its form, sentences, and choice of words.

One conclusion that may be drawn from what we have said hitherto is that there are no neutral texts. Those texts that appear to be impersonal, but definitely not neutral. Since all texts have human origins, they are always influenced by both conscious and unconscious thought processes. One may choose to give an impersonal account for different reasons, such as the demands of convention, or simply in order to mystify the reader under pretence of being neutral, factual and trustworthy.

2.3.2.7 The visible side

In the relation between thought and language — the way in which the conscious and unconscious aspects of thought become visible — there exists not only the side that is invisible and elusive, but also one that is clearly visible. It consist of all the countless ways and means we have of writing down our ideas and thoughts. But whatever we do, whatever routines and habits each of us has, these procedures are at the same time directly related to our thought processes.

Let us take an example from school. Some teachers in primary and secondary school decided to work with their children so that they would have a chance to prepare themselves mentally before writing. The work started on the Monday by stimulating ideas about what they should write on the following Friday. After each child had decided on a subject, the work of developing these ideas started. This could consist of discussions on the subject, role play, expressing it in pictures, talking about it at home, carrying one's "thinking book" around with one in order to jot down ideas. After a few days, they began to work with the structure, either by writing headings and ordering them and/or by drawing a series of pictures. Not before this did the writing begin. This preparation work had a very favourable effect on all the children's writing in all cases. They wrote more than usual and showed great enthusiasm; their stories had a main theme and were formally and linguistically better than previously.

In this way the children are helped to find means of making their thoughts visible. Each such disclosure of their thoughts, through any of these different forms of expression, leads to deeper thought processing. A process is going on in which the mutuality between thought itself and the expression of the thought emerges, which is the very basis of development of a theme into written language.

This example may not seem especially remarkable or unique. This kind of preparation is made by everybody writing for others. It may even be so self-evident that one hardly thinks of it. The remarkable thing about the example is that it has to do with children and that it is taken from school. Everyone who has gone through school knows that writing essays is not usually like that.

There is hardly an adult who, directly after having an idea, sits down and immediately produces a finished text. And if they do exist, they are extremely rare. However, it may be that those who consider that they function like this do not count the invisible preparation process in writing. They do not, like the majority of some other people, jot down a few words or sentences so as to remember a thought structure, but work with the idea completely "inside their heads". When they subsequently begin to write, it flows smoothly.

The variations in how this thinking-writing process functions are as many as the number of individuals involved. Some people make notes or tables of contents, others make a sketch or a model, and still others write maxims or key sentences or combine different ways of structuring their thoughts. But each person does it in his own personal way.

One could regard this as a self-regulating process. The pre-comprehension that comes to the fore by the idea may be more or less clear and more or less comprehensive. Let us suppose that the idea is vague (pre-comprehension frame). One ponders a little (thought-processing frame), writes down a few words or makes a sketch (writing frame). One has formulated a message — at this stage directed to oneself (written message frame). This extends one's pre-comprehension (the arrow to pre-comprehension), one's thoughts become more lucid (thought-processing frame), and one writes another few words or sentences (writing frame), and so on. In this way it can continue layer after layer, until eventually it will become something coherent — a paragraph or a page or two — that also will provide a good pre-comprehension for one's continued writing.

That this process regulates itself is obvious when one considers what the "frames" in the model contain. How one works depends entirely on one's knowledge and experience in all respects: the choice of subject, aims, and so on. A writer who has considerable experience can work faster and more effectively than someone who writes less often. The skilled writer has worked out viable routines; he is accustomed to and is skilful at concentrating his thoughts, he may perhaps write down complete sentences directly and can more easily calculate how long it will take to write a text.

2.3.2.8 Who can write?

A common assumption, or rather misconception, is that the ability to write is a question solely of talent — some people can write and have this gift, others not. Seen from the outside, this assumption may appear to be reasonable and understandable. But that means that one regards the ability to write as something isolated and completely independent of the individual's life history.

Let us take Pelle and Britta as an example. Both of them go to a senior school in the same class. Pelle has so-called problems in reading and writing; he reads only when he has to and then only by making a great effort; he forces himself to write a couple of lines when essay writing, with bad handwriting and bad spelling. Britta, on the other hand, is long past the book-devouring stage. She loves to read and reads a good deal. When she writes it is as though the words flow from her pen — everything correctly spelt. Here we have Pelle who cannot write and Britta who can, and it would be an easy way out to state how talented Britta is and how untalented Pelle is. But the picture is completely different if we go back and look at their respective life histories.

Pelle is an only child. His mother had problems in reading and writing and Pelle has never seen her reading a book. She hardly ever read stories to him when he was small, and Pelle's father never had the time. What Pelle has seen of stories and children's books has been through the television and nursery. His mother is, and always has been, reticent. When Pelle began to talk rather late, he was confronted with monosyllabic answers, hums or actions without words, that is, if he pointed and said something that resembled a word, he got what he wanted but it was never confirmed in words and sentences. For Pelle, this meant a limited vocabulary and pronunciation that for a long while was childish. During his stay at the nursery, he improved somewhat. When, like all other four-year-olds, he began to ask questions, he was confronted with "Be quiet!", "Don't ask so much!", "Stop it!". Or when he tried to express his own conceptions of the world around him, or ask existential questions of life and death, such as where everything comes from and so on, he was given responses like "Wrong!", "Fantasies!", "You wouldn't understand!". At the nursery, no-one had the time or the opportunity to deal with these questions. So few adults for so many children does not give each child the chance for talk that he needs on these matters.

For Pelle, this has meant that he does not possess the words and expressions for many aspects of life; that he has interpreted his situation as meaning that it is best to keep quiet, since what he says and thinks is wrong anyway; that there is no point in formulating his ideas since no-one takes any notice, is interested, or has the time. This was the situation when Pelle started school, but even there he did not have a chance. He was already "knocked out" at the infant school stage, since the school lessons were dominated by text-book teaching, that is, there was no opportunity for thinking and understanding; abstract incomprehensible texts with exercises to fill in; Swedish language based on technique and form, and so on. Pelle learned to read, technically speaking, but he got nothing out of it. He has actually never seen the point of being able to read. Throughout his school days he has sat through essay-writing lessons and suffered defeat, time and time again. Like many others he could not understand how to go about "making something up" or what "making something up" had to do with spelling, handwriting, real sentences and grammar. Some could always manage it, especially Britta, but not Pelle and many others like him.

Pelle and Britta have been in the same class throughout this period. They have thus been given the same teaching. But Britta's experiences, both before starting school and, later, outside school, are completely different. From birth she has been part of a rich linguistic interplay with her parents, sisters and brothers, and later also with other people. She was, therefore, able to develop her language with a rich vocabulary that comprises phenomena both from the world around her and from her inner life. Her questions have always been taken seriously; have been responded to with interest, and have been answered, and so have her models of explanation. She has even been encouraged to talk about how she thinks everything fits together. She has thus developed a firm self-confidence and a confidence in her linguistic and intellectual abilities. The teaching has not completely prevented this development, only in certain respects and in connection with certain subjects.

Pelle's problems are caused, among other things, by the fact that he has never been given the chance to develop an inner speech — linguistic thought. The basis of this is the social and linguistic interplay between the child and an adult from birth. It is thus a question of dialogue, of mutual exchange of ideas and respect; questions and answers; juggling with opinions and hypotheses. This social dialogue is a kind of model for the inner dialogue, but only on condition that the child, within the social interplay, has experienced that he has something to contribute and that his thoughts and wonderings will be responded to and are of value. In this social interplay the spoken language is developed — our most important means of orienting ourselves both in the world around us and in our inner selves — and from it our inner speech is developed. Pelle's inner world remains dumb. It is not that he cannot think, but that he does not know what he is thinking. Since he has no inner speech and hence no inner dialogue, he has no means of giving voice to and expressing his thoughts linguistically, either verbally or in writing. He can manage to perform some tasks mechanically: "Do it like this", and he can do it. But as soon as the job demands reflection and consideration, he cannot manage it. He simply does not understand what "think about" means, in the same way as he does not understand "make up something to write", since he does not have the experience in doing these things. In order to develop an inner speech he must first experience its external social equivalent. (See further book 2, chapter 2).

If we now look at Pelle's and Britta's ability to write in the light of each one's knowledge and experiences, not least of which are those related to thought and language, it is not reasonable to regard the difference in their ability solely as a question of talent. Classification according to talent becomes an absolute determinant or a self-fulfilling prophesy. There is no evidence to suggest that Pelle, in another class-room situation, could not have developed an ability to read and write comparable to Britta's, nor that, in different circumstances, he may still have a chance to develop it. The sad thing, however, is that the later it gets, the harder it will be to change things. The negative identity, that comes from not being able to succeed, from not counting and from not being of any value, that Pelle has built up round himself, will be more and more confirmed. He is continually getting evidence that this identity is the correct one, and it will become more and more difficult to reverse this trend. His own conviction that he is untalented and stupid is confirmed by an equivalent belief in those around him. What we have now said does not mean that we deny that people are born with different genetic equipment, but we mean that how this genetic equipment finds expression depends to an enormous extent upon the individual's interplay with the world around him.

2.3.3 TO SUM IT UP....

Writing is thus a process of unconscious and conscious, simultaneous and gradually-occurring events. The writing process includes both the creation of written message and the revising of it. When revising the message, one reads through it, which is necessary for one's pre-comprehension before continuing to write. Otherwise one would lose the thread. It is hard to remember, not the rough outline of what one has written previously, but the details of how one formulated the last three pages, or last three paragraphs. (The revising of written language is illustrated by a model that combines the reading and the writing process.)

Writing is thus not just the ability to put some words down on paper in a linguistically correct way. To formulate a message is a complex process that requires room for thought in order to come into being; that requires to be thoroughly revised if it is to give the readers of the text, if not an enjoyable, then at least a fluent, stimulating, and lucid reading experience.

REFERENCES

Lo-Johansson I (1981) *Att skriva en roman.* Stockholm: Liber Forlag.
Vygotsky L S (1969) *Thought and Language.* Massachusetts: The M.I.T.Press.
Vygotsky L S (1981) *Psykologi och dialektik.* Stockholm: P A Norstedt & Soner.

2.4 THE READING PROCESS

by

Birgita ALLARD and Bo SUNDBLAD
Institute of Education, Stockholm

2.4.1 SUMMARY

In this paper, we shall try to show that the idea of what reading is, or how it can be defined, is a very important question. But in order to be able to deal with this question it is also important to be aware of the difference between reading as a *phenomenon* on the one hand — as it appears and indisputably exists — and models of explanation or *theories* of reading on the other hand. The existing phenomenon of reading can be explained and defined in more than one way. In other words, *for this one and the same phenomenon there can exist more than one model of explanation or theory.* A theory is nothing more or less than a series of logical consequences drawn from data and from assumptions about the phenomenon in question.

There is still another important factor to draw attention to. In education and in teaching, all the methods and procedures used are based on ideas, assumptions or theories. *No pedagogical practice is free from theory,* whether or not the educator or teacher is aware of it. The idea or conception of reading has a direct influence on the construction of methods for teaching reading. These methods are nothing other than the logical consequences of some theoretical standpoints — invisible or visible, unformulated or formulated. But many teachers believe that the methods they use have nothing to do with theory. This separation between theory and practice — experienced by teachers as the non-existence of theory — is, among other things, a result of teacher-training, since methods are usually taught with no reference to their underlying theory and this too often leads to the confusion between theory and the phenomenon.

Different theoretical assumptions or models of explanation have different practical and methodological implications. Since theory belongs to the area of research, we can also put it in this way: the choice of a research approach and of research methods is implicitly or explicitly based on a definition of reading. Thus a seemingly objective description of different behavioural parts which can be observed is influenced and governed by ideas and theoretical assumptions of reading as a phenomenon. The definition of reading underlying research has a direct influence on reading instruction in school.

2.4.2 DEFINITIONS OF READING

2.4.2.1 *The definition of reading varies with different scientific approaches*

Questions concerning what is scientific and what is not, and the distinction between science and non-science are topics within the field of theory of science or the philosophy of science. During the 20th century two main general schools of metascience can be mentioned with relevance for reading. One goes back to the 1920s and is called logical empiricism or neo-positivism. In short, the ambition was to create a "unity science" where all disciplines (including the behavioural and social sciences) should be seen as branches of the natural sciences with physics as the scientific ideal. The aim was to find strict scientific criteria in order to discard metaphysics. Any statement which does not refer directly to experience and the observable world is metaphysical and thus non-scientific. This meta-scientific school gave legitimacy to the behaviouristic movement. When the principle of operationism was developed by Bridgman in 1927, in the realm of physics, it was in accordance with this scientific ideal. To illustrate what is meant by operationism, Bridgman uses the concept of length...

> "...what do we mean by the length of an object?/.../To find the length of an object, we have to perform certain physical operations. The concept of length is therefore fixed when the operations by which length is measured are fixed: that is, the concept of length involves as much as and nothing more than a set of operations; *the concept is synonymous with the corresponding set of operations* (1927, p.5)

Applied to reading, this concept can be defined by the operations used to measure it. Reading is what is measured by a reading test. Consequently, the definition of reading changes from situation to situation depending on the construction of the test or measurement. But the observable parts of the reading behaviour must be identified and made measurable. Reading tests used for diagnosis and for scientific purposes define the phenomenon of reading and will thus lay down the rules for how reading has to be taught and learnt.

All scientific disciplines which have all aspects of the human being as the subject for research can either adhere to the above-mentioned school or to another school of metascience. The latter includes many subschools or metatheoretical approaches but they can be put under the heading of "human sciences" since they do not have the natural sciences as the scientific ideal. Human qualities (including social interaction) which are not directly observable are not rejected as metaphysics but can be studied for themselves and must not be transformed into their observable counterparts. From this viewpoint, reading is not exclusively seen as a behavioural repertoire, but is viewed as a linguistic phenomenon, in which human beings are involved in communication. Written language as a formal system is not the main focus but rather its cultural, social and individual functions. Learning to read and write, and the further development of these abilities, is part of language development, which starts as early as birth. The preconditions of language development are the interaction between the child and environment — its

156

qualities, quantity, contents and meanings. Language development is also related to other mental qualities of the individual, seen as interaction.

During the developmental process from birth into adulthood the child is actively constructing concepts and ideas of the world, including himself. The child's conceptions of language can be traced in his way of using it, exploring it, inventing words, etc. Long before a child starts to use words, he masters the functions of speech (the communicative aspects). Long before a child can read and write in the conventional sense, he has grasped some of the functions of written language (if he has experienced its functions). He "pretends" to read — though from his own point of view he really is reading — and from his intonation and attempts to "formalize" his words he shows that he has an idea about the differences between spoken and written language. He writes, but not by using the conventional symbols, or if he does, he does not use them in the regular ways. He recognizes and draws letters. Much can be learnt from a child's spontaneous explorations of the written language, provided that we are not too biased by the way in which reading and writing are usually taught in school. When Josefin was about 4 or 5 years old, a TV series aroused her interest in letters. After the second programme she told Bo that now she knew all the letters. In front of a wall chart with all the alphabet, she said: "I know this one, and this one...", and so on, pointing at the letters one at a time. She was very proud and self-confident. When she had finished, Bo pointed at "F" and said: "Okay, and what's the name of this one?" She answered: "Oh, I don't know, I have forgotten it, but I *know* it"!. Bo was just about to comment on that, but he swallowed his reply. He intended to say: "But then you don't know it!" Knowing letters usually means being able to name them, so we had to revise our view.

2.4.3 LANGUAGE DEVELOPMENT

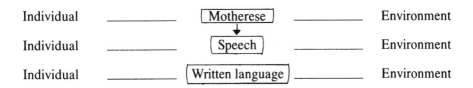

Figure 1

In Figure 1 we illustrate language development. We use the word "motherese" in order to focus on the social and communicative interaction. It is mainly a non-verbal language, a communication with the body, gestures, mimicry, the eyes and on the part of the child, accompanied by sounds and later, when speech comes into the picture, by words and then also intonation. The mother also uses words in this communication. Motherese is the basis for speech development, which in turn is the basis for reading and writing development, or, to quote Vygotsky:

"Understanding of written language is first effected through spoken language, but gradually this path is curtailed and spoken language disappears as the intermediate link." (1979, p.116)

2.4.4 THE READING PROCESS

Let us turn back to the concept of reading. When we are reading we are not aware of what we are doing. Most of the time our attention and awareness is directed to the meaning of the text. The reading process is to a certain extent an unconscious process. Still what this process comprises is important. Anyone who can read can, after some moments of reflection, enumerate some of its parts. But still more important is to recognize how these factors are inter-related and interact with each other. The reading process, presented in figure 2, is *one* model of explanation[1]. It encompasses the main factors which are at work when a person is reading.

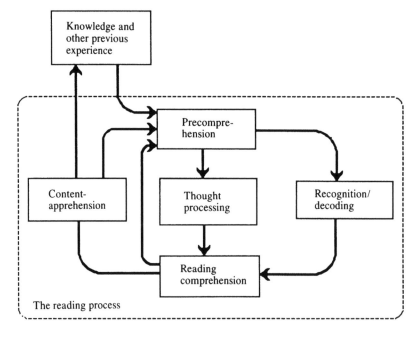

Figure 2

The big square refers to the actual reading process. The outer square summarizes all that we are, all our experiences and all that we know and have learnt, independent of whether we read or not. Here is also included our linguistic

---

[1] 1979 Edfeldt presented his model for the reading process (in his book 1982). This is our interpretation of his model.

experiences and competence in all senses. Our knowledge and experience, not least our reading experience, are of great significance in all reading situations.

Whenever we are confronted with a written passage, whatever it may be — a book, a newspaper, a letter, a note, a sign, or the subtitles on the television screen — a process is started at an unconscious level. We so to speak activate, from our knowledge and experience, all the aspects we find relevant in connection with what is written in front of us. As little as parts of a word, or one or a few more words starts this process. If a text begins with "Once upon a time..." we all know what to expect, not only with respect to the contents but also to its structure, language, and so on. This process gives what we call *precomprehension.*

2.4.4.1 Precomprehension

Precomprehension directs and guides decoding. While decoding we simultaneously elaborate the contents of precomprehension – thought-processing. Decoding and thought-processing leads to reading comprehension. If we exclude decoding, it is no longer a question of reading. We have only been thinking. If we take away

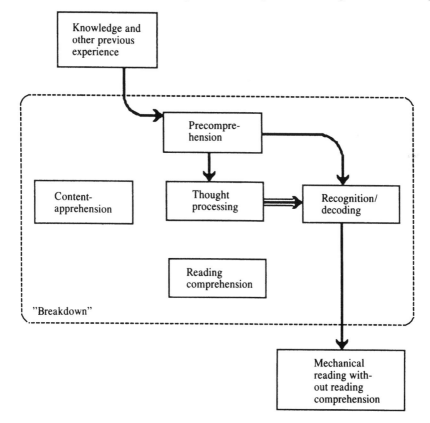

Figure 3

159

thought-processing we are left only with decoding, which can hardly be called reading since it does not lead to comprehension. We can call it mechanical reading, illustrated in figure 3. You surely recognize the following situation. It is late in the evening and you remember that you have to read a paper in order to be prepared for a conference the next morning. You start reading it and after three pages you stop and realize that you have not the slightest idea of what you have read. You have decoded, but where were your thoughts? If this happens too often, it is easy to panic. Some people may think that they are bad readers.

After a word or two, or after a sentence or a paragraph, precomprehension is enriched (the arrow from reading comprehension to precomprehension). Precomprehension is continually enriched and elaborated during the ongoing reading process. The more we read of a text, the better precomprehension becomes. Since precomprehension directs decoding, reading becomes more fluent. If we, from the start, are acquainted with or have a good knowledge of the topic we are reading about, we have a good and rich precomprehension, and hence we read the text fluently and rapidly. If the subject is unknown to us, or the language is very abstract with complicated sentences, pre-comprehension is poor and reading the text will be arduous. Now and then the process stops. We have to re-read words or sentences and sometimes we must look words up in a dictionary. Reading speed is very slow. Reading speed is always a function of precomprehension.

When reading a novel, the first chapter usually takes a longer time to read than the following ones. A common expression is that it takes time to get into the book. We have to learn about the characters, the milieu, and get used to the author's style and use of language. It can be illustrated as in figure 4.

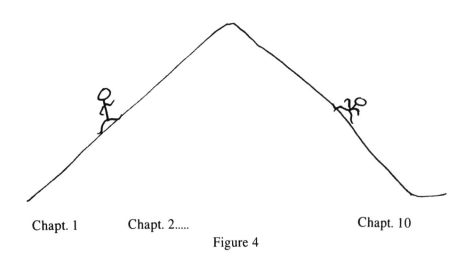

Chapt. 1   Chapt. 2.....   Chapt. 10

Figure 4

First we have to go uphill (chapter 1), and then skiing downhill is pleasant and enjoyable. Some books are like climbing a steep, high mountain. To children, with little reading experience, many reading situations may appear in this way. Why not give them a lift up to the top, so they may have a chance to enjoy the journey downhill? That means helping them to acquire a good precomprehension, before they start or go on reading. It can be done either by reading a chapter or two for them or by telling them about the story. This is, of course, extremely important when they have to read texts of facts in any school subject. With poor, or no, or wrong precomprehension it is sometimes impossible for them to read a text, not only to understand what they are reading, but also to decode the unknown words. It is not unusual for them to change the unknown word into a similar but known word.

Another difficult situation for children is when they have to read separate words which are not embedded in meaningful sentences. Pre- comprehension is always at work. After having read the first word, precomprehension directs decoding of the next, and so on. Let us take the following words as an example. Although they are in Swedish you may catch the idea:

fromage, bandage, komage

All adults reading these words make the same mistake. All the words end with "age", but only the first two words have the same pro-nunciation. (The last means the stomach of a cow.) Everyone pronounces the last word in the same way as the two preceding ones. This means that in all kinds of tests where children have to read separate words they are never able to perform in accordance with their actual ability. Some children may experience this as a failure. Since many tests contain this kind of meaningless decoding, the test itself can lead the child into mechanical reading. If we add other "impossible" reading tasks, they, together with the test, may give rise to what is called illiteracy.

2.4.5 CONTENT APPREHENSION

Let us go back to the model (Figure 2). What we have said about "getting into a book or a subject" also concerns what we call content apprehension. We always create whole structures or "pictures" of the content, irrespective of how little or how much we have read. During the reading process we may, perhaps, have to reconstruct the "picture" which, for instance, often happens when reading detective stories.

We also learn and gain experience from reading (the arrow from content apprehension to knowledge and experience).

Whenever the process is underway and is working smoothly, we can say that it is an indication of a good reading ability. This is as equally valid for a 7-year-old as for a 27-year-old, etc. Saying that a seven year old is a good reader and saying the same of a twenty seven year old may seem like a paradox, since there must be big differences between them. However, against the background of their respective qualifications it is still true. They differ in respect of knowledge and experience and hence in respect of precomprehension. The 7-year-old is not as advanced as the 17-year-old, and can only manage to read texts suitable for his or her level of knowledge and experience. But still he or she can do it fluently. If the same child is forced to read a text above his level of knowledge and with an unfamiliar syntactical structure, he does not read it fluently and the process very often stops. Unfortunately this is often taken as an indication of poor reading, which is not the case.

2.4.6 CONCLUSION

There are many conclusions to be drawn from this model, not least concerning learning to read. In short, the main principle is that children must be given the opportunity fully to use their knowledge and experience with reference both to the contents of the texts and to their linguistic ability and experience. Most primers are based on the learning of decoding skill initially, and hence the separate words and sentences with "easy" words (easy only in relation to decoding and not to meaning) for training children to decode do not make much sense to the children. In order to become a literate person, reading must, from the very beginning, be an exciting and meaningful experience. We mean that besides a number of cultural and/or social factors this "learning-to-read" situation is an important factor behind the "new illiteracy".

REFERENCES

Bridgman P W (1927) from Schultz D P (1969) *A History of Modern Psychology.* Academic Press, New York.

Edfeldt A W (1982) *Lasprocessen - Grundbok om lasforskning.* Liber Utbildningsforlaget, Stockholm.

Vygotsky L S (1979) The Prehistory of Written Language, in *Mind in Society.* Harvard University Press.

2.5 THE CHILD'S CONCEPTION OF READING

by

Lars-Erik OLSSON and Gosta DAHLGREN
Department of Education and Educational Research, University of Goteborg

2.5.1 SUMMARY

This paper reports a study of reading in general, and reading instruction for beginners in particular. The perspective is that of pre-school children (5-6 years old) and first-grade children (7-8 years old). We have used in-depth interviews with 80 children to describe readers' and non-readers' conceptions of the reading phenomenon.

2.5.2 BACKGROUND TO OUR RESEARCH PROBLEM

Reading and writing abilities are essential skills in a modern society. These basic skills have recently been discussed in Sweden in terms of "the quality of reading/writing performance" and "methods for early reading instruction". The focus of these two discussions is the function and the structure (form) of reading.

The reading ability of schoolchildren was discussed in the seventies, and the term "functional illiteracy" was a key term in this debate. Alongside this discussion of the general performance level in reading and writing, there was also a debate concerning reading instruction for beginners. Opponents to the so-called "old" synthetic reading methods heavily based on phonics (see Chall's (1967) definition of the term, p.149) argued for a method based on the child's own vocabulary. This "new" method for beginners was labelled analytical and had similarities with the learning experience approach (LEA). It is interesting to note the difference between the Swedish debate and the "great reading debate" in the USA in the fifties and sixties.

The problem of tackling compulsory education in Sweden interested laymen, teachers and researchers. One of the research paradigms in our department offered a promising perspective in which to formulate our research problem and investigation of reading. The research paradigm is called the "second-order perspective" and centres on how people describe different aspects of their surrounding world. The label of the approach is phenomengraphy (Marton, 1981). The focus of our study is to describe children's conceptions of reading.

2.5.2.1 The research questions

The study focuses on two main questions: what conceptions do children have about the usefulness of reading and of the reading process? The first regards the WHY-question (function) and the second the HOW- question (form) of reading. Both questions taken together can be philosophically, theoretically and empirically integrated in the superordinate question: what is reading?

2.5.2.2 The child's cognitive prerequisites for acquiring our written language

We present here the discussion between Piaget and Vygotsky about the child's early egocentrism and egocentric speech. Egocentric speech is a phenomenon described by Piaget and discussed by Vygotsky. Vygotsky regards egocentric speech as a precursor to what he calls "verbal inner speech" and pure thought (thinking in meaning without identifiable form). Piaget and Vygotsky both identify egocentric speech but draw different conclusions about its origin and functional and structural development.

Vygotsky disagrees with Piaget's assumption that the child's egocentric speech is a symptom of inadequate social adjustment and that the overcoming of this social egocentrism is indicated by a gradual disappearance of the child's egocentric speech.

Vygotsky suggests a different interpretation. The small child is trying to internalise verbal behaviour previously used for direct communication purposes. Egocentric speech can, in his opinion, unveil the development of inner verbal thought. Both the structure and the function of this egocentric speech within such a frame of reference is quite different. The child is directing his speech to himself and not to others. Gradually, this language develops into verbal inner speech.

The child's ability to assume the role of others is a central issue in reading and writing. The reader's ability to understand the writer as a person "behind" and "in" the text, and the writer's ability to understand the reader's perspective, accentuate the role of social egocentrism in thinking. Studying inner speech via egocentric (oral) speech makes it possible to study indications of the function and structure of inner speech. Inner speech and written language handle subject and predicate in different ways. Written language has fully developed subjects and objects, while inner speech lacks these aspects. Egocentric speech and inner speech are fragmentary, emphasising the predicate in a different way. Therefore, they are termed "predicative language". In this respect, written language and inner speech are quite different from each other. The developing subjectivity of inner speech can be described by the statement "Intentions taking over at the cost of structure".

These central aspects of the child's "different" languages are vital to the researcher's understanding of the child's comprehension of reading and writing. He or she rapidly develops an inner verbal speech during the period of 5-8 years of age. The cognitive process of observing the differences, similarities and possibilities of the "languages" is of importance when learning to read and write.

2.5.2.3 Earlier reading research in the second-order perspective

We have concentrated on studies of how children describe and conceptualise the reading function and the reading process, as our experiential aim is to describe children's conceptions of the phenomenon of reading.

Research on reading in the second-order perspective has two relatively distinct and unrelated traditions, one in England and the other in the USA. The English branch started in 1958 with an article by Reid and was later followed by studies by Downing (1969) and Francis (1973). One characteristic of these studies is that they are based on intense interviews, observations and performance testing of a small number of subjects.

The American tradition started with a study by Denny and Weintraub (1963) and was followed by Johns (1971). One special feature of this research is its survey character, with short interviews with a few central questions administered to a large number of children. In one interview study, 1,655 children from grade 1 to grade 8 participated.

The results of these studies suggest that pre-school children have vague expectancies of what reading is going to be like, what the reading act contains, and what use you can make of it. They lack relevant vocabulary for talking about this activity and they are cognitively confused.

The English results are delightfully detailed in presenting the children's different conceptions of reading. Our most quoted source is Francis' (1982) study of 10 children who were observed during their first three years at school.

2.5.3 THE RESEARCH PROBLEM AND THE LINGUISTIC TERMINO-LOGY

If you ask a person to describe a language act, he is likely to reflect on his own cognition. Metacognition related to language produces a metalanguage (a set of linguistic terms used for talking about language). In learning to read, children face the difficulty of understanding the content of the so-called "reading instruction register" (Downing, 1976). We find support for the idea that learning written language produces awareness of spoken language as well. This may, in turn, create awareness of the thinking process and thereby give the child intellectual self-control.

Language competence and language awareness are two different kinds of data levels. The first relates to language used in ordinary life situations and the second to language as an object for reflection. We use a "strong definition" of language awareness as the child is asked to express awareness verbally. This definition of reading awareness is theoretically and empirically related to the child's reading performance.

2.5.3.1 Method

Within a second-order perspective, the researcher's aim is to describe the world as perceived by people. Results are presented in categories describing people's conceptions of a certain phenomenon. Phenomenology and phenomenography are related in their search for the world as it is perceived, but they differ in their analysis of data and presentation of results. Phenomenology is directed towards a description of the true invariant meaning of a certain phenomenon, while phenomenography describes people's conceptions of a phenomenon in qualitatively different categories. The difference between phenomenology and phenomenography has consequences for the way results can be used in educational practice.

When collecting our data, we used Piaget's (1977a) clinical interview method and its criteria for judging the relevance and the reliability of interview data. Piaget adds two further ways of expanding the knowledge of a child's conception of a phenomenon. One is direct observation of actions and the other is an analysis of children's questions. In our study we make observations on the children's reading "actions".

Vygotsky (1982) is critical of the Piagetian aim to look for the attitude of spontaneous thought and the uninfluenced mind of the child. The Piagetian aim partly explains why the developmental stage theory has been difficult to adapt to educational practice for subjects other than maths and natural sciences. Vygotsky stresses this difficulty especially for phenomena with a weak structure and an ideological and pluralistic theory base. A context-free description of thinking may be a research goal *per se,* but it is futile for many didactical purposes. The Piagetian problem of "decalage horizontal" seems to Vygotsky to be a distinction between scientific concepts and spontaneous concepts. A person's scientific concepts become conscious within a system, while spontaneous concepts refer to actions and awareness necessary in everyday life. *Decalage horizontal* refers to the difficulty of solving logically identical problems in different contexts. This difficulty can largely be explained by referring to the relation between everyday thinking and scientific thinking. *Decalage horizontal* as a stage-defined developmental problem in thinking is, for Vygotsky, not an abstruse research question but merely a contextual and didactical one.

The context-dependent variation in answers relates to the educational level of the subject, the ability to understand what is demanded by the researcher, the conception of language and the knowledge of and ability to work through a strictly hypothetical discussion including nothing but the given premises. One way of labelling some of these factors is to use the term "interview contract"

166

(Rommetveit, 1977). The complete understanding of such a contract is a description of the context-dependence of a subject's answers.

2.5.3.2 *The research design*

Our data consist of 133 tape-recorded individual interviews with 80 children. The children came from seven different pre-schools in Gothenburg (the second largest city in Sweden, with a population of about 600,000). The participating schools were chosen from the four districts that are administered by the municipal social services. The group consists of approximately the same number of boys and girls. Without being randomly selected, the group represents a variety of home backgrounds, ensuring differences in conceptions of reading. A follow-up study was administered with 53 of the oldest pre-schoolers after one year in school (school starters in Sweden are seven years old).

The pre-school study concentrates on questions pertaining to the function and form of reading. The two main questions can be exemplified as follows:

> What reasons do you find for reading/learning to read? How is reading done?

The questions were systematically reformulated in a number of different ways during the interview. The conversation began as an open interview and focused gradually on reading in an "adult manner". The child played with a letter puzzle, wrote a few words, the interviewer read a little story and finally asked some specific questions at the end of the interview. We present here some of the questions in the interview, just to give the reader an idea of the *type* of questions asked.

> Can you read?
> What can reading be useful for?
> How is reading done?
> What must you do to learn how to read?
> When will you learn to read?

After asking this, we read aloud from a book that the child himself chose from the pre-school "library". Then we asked the questions again.

We also asked the child "where" and "what" he/she read in books (texts, pictures, letters, numbers, words) and how to write names and short words with a pencil and a letter puzzle. One part focused on the relation between a real object, a picture of the object, the name of the object (spoken and written) and the child's explanations of the relations between these different forms of representation. We studied the child's conceptions of the writing convention (writing direction, words as units, the alphabetic construction and its basic idea). Finally, we asked some questions appealing to the child's imagination, such as, "What would happen if nobody could read?", "Could the world run out of letters?" and "What is the origin of reading and writing?"

The school interviews (53 children out of 61 possible school starters were studied a second time at the end of grade 1) concentrated on our two main questions. The child brought the textbook used in the classroom to the interview. Forty-nine out of 53 children had the same primary reader. The other four children had a "language experience approach" in their first reading instruction. Tests of reading performance were administered. We used one speed test measuring reading speed and type of reading errors, and another test measuring vocabulary and reading comprehension (all tests were standardised and intended for Swedish grade 1 children at the end of their first year at school).

2.5.3.3 The analysis of the interview protocols

The goal for the analysis of the interviews was to describe the child's conception of reading. The descriptions are presented in qualitatively different categories. The term "qualitative" does not refer to a classification into "good" or "bad" interview answers; instead we try to describe different conceptions even if they should be wrong, mythical and biased according to various widely accepted definitions of reading. Our work is entirely focused on experiential descriptions.

The interviews were written down on paper, word by word, and afterwards corrected for listening errors. The first step in our work was to identify interview parts corresponding to the how- and why-questions respectively. By working out tentative schemes for category descriptions, the comprehensibility of children's conceptions was successively deepened. The two final category systems were tested for reliability by letting another researcher analyse and categorise (the reading function and the reading form respectively) every child. To do this reliability test, the co-judge was allowed to use only the description of the categories without taking any example directly from the interviews.

2.5.4 RESULTS

2.5.4.1 Our answers and the school context

If we ask pre-school children about the content of school work in grade 1, almost every child mentions reading, writing and maths in that order and importance. Children describe reading as closely knitted to reading instruction and school work under the guidance of professional teachers. Children also mix learning in general with learning to read.

2.5.4.2 The function of reading

The children's descriptions of the reading function can be classified into two qualitatively different categories. A: reading as a possibility, and B: reading as a demand. We will first present the sub-categories of A.

2.5.4.2.1 A Reading as a possibility

A1 The child describes reading of different texts as a motive for reading.

The motive has a surface aspect of the function of reading. The child mentions different kinds of textual media such as books, letters, newspapers, slips of paper, TV-texts, signs, labels and game instructions. The answers in the category focus on different reading media but include no notions of reading content.

A2 Reading content as a possibility

The child's motive is the content of what it reads. This use of reading contains descriptions of A2(a): experiences, A2(b): information and A2(c): knowledge acquired through reading. Reading and writing are also A2(d): means to communicate with other people.

A2 implies category A1, but in this category the child includes what a reader can do and wants to do right here and now. Immediate usefulness is in the child's focus.

A3 Reading competence as a possibility

The child describes reading competence as a means for other ends. This use implies the primary use (category A2) but stresses different ways of reaching other goals by means of reading competence. This use of reading competence contains descriptions of A3(a): independence by competence, A3(b): usefulness in school and A3(c): usefulness in adulthood.

Answers in category A describe reading in terms of its usefulness. The child explains the reading function by stressing personal usefulness as the most important motive.

The three main A-categories indicate a developmental order. The first describes texts as external objects but disregards the content of what has been read. The second category stresses the content and its immediate usefulness for a reader. Category three implies the previous two, and points out the usefulness of reading competence for reaching goals other than just the content of what has been read.

2.5.4.2.2 B Reading as a demand

A qualitatively different conception of reading is presented by children in category B: reading as a demand. This category has the same structure as category A but differs in emptional tone. The usefulness of reading is pressed upon the child/demanded by others. Category B reflects the sub-categories of A3.

B(a) "You must be able to read in order not to bother others". In this category you find a wish for independence, but compared to category A3(a) this wish is verbalised as something pressed upon the child by people in its environment. The wish is not expressed by the child but by others, and it is not fully internalised.

B(b) "You must be able to read, otherwise you will be teased in the peer group". The child expressed an emotional feeling as an illiterate of not being fully accepted. This motive is strongly knitted to the self-concept of the child and describes one important aspect of the self-image in the 6-7 year span.

B(c) "You must be able to read in school". The child describes that one demand characteristic of schools is their institutionalised right to expect pupils to be able to read and write. We find different conceptions of this, and we can define them as myths of reading and schooling. One myth is that you have to read before school entrance, otherwise you will not be accepted as a pupil. Another detrimental conception is that you must not read before starting school. You have to make sure you learn how to read in the right milieu and under the guidance of professionals. A third way in which the child conceptualises school and reading is that you will be expelled if you cannot read. Non-readers have a short time of trial to demonstrate their reading ability, and, if not successful, they will be expelled.

In category B the child describes how to avoid "threats" by learning to read. The category focuses on the emotional tone subsumed in the expression "reading as a required competence!. The child has difficulties in overlooking this emptional stumbling block and expresses the usefulness of reading for himself.

Anxiety related to reading, reading instruction and reading performance is strongly linked to school entrance. When we talked about the content of school work with children, they pointed out reading as a very central school activity, that provokes these emotional reactions. Not knowing the "nature of reading" leaves the child with a feeling of insecurity, expressed while talking about the usefulness of the reading activity.

The above system of categories concerns the children in pre- school and in grade 1. For grade 1 there will be a third main category added to the system of categories of reading function already presented.

2.5.4.2.3 C Improving the reading competence as a motive for reading

The child describes that learning to read, and to improve this competence, is a motive in itself.

Besides the qualitative differences between pre-school and grade 1 answers there are also differences in distribution within the system of categories. Already in pre-school 40% of the children described reading and writing as communication acts. They grasped the functional aspect before they had the actual reading competence. It is remarkable that children in grade 1 give fewer answers of this kind in their descriptions of reading. Reading competence described as a demand from "others" dominates on the other hand, among the answers from the children who have not yet started school. They practically disappear in grade 1.

2.5.4.3 The form of reading

We introduce the descriptions of the reading act by presenting results about children's terminological difficulties concerning the reading instruction register.

Children's understanding of the meaning of the reading terms indicate difficulties in isolating and relating the concepts letter, name, alphabet, number, word and sentence. Names are mixed with letters and are not identified as special words. Letters are numbers and vice versa. The alphabet contains both letters and numbers. Counting and reading are defined as undifferentiated symbolical activities.

Another way of introducing children's knowledge of the reading act is by analysing and describing their explanations of the relation between pictures and text. Children's preferences for text information are verbalised in three different ways. The first suggests that pictures are made unambiguous by the use of text information. The second implies that text information directly tells you what people say, think and act. Text information is also superior to information given by pictures because of its greater importance and usefulness in our society. The child expressed its preference for pictures by saying that pictures are more real and easier to interpret than text information.

2.5.4.4 The child's conception of the reading process

The content of the reading process is described in four main categories – A, B, C and D.

2.5.4.4.1 A Context-related process of reading

The child describes the reading process as focused on "things outside or accompanying" the text. The precise meaning of the term "context" within this category refers literally to the translation of *con textus* which means something "following the text". The context category consists of five sub-categories.

A1 The act of reading is described as an external procedure identified by its behavioural characteristics.

This conception of reading describes the reader as a person talking, sitting, holding a book, looking at and turning over pages. Texts are presented as objects for these activities but the content of the reading act is not mentioned.

A2 Reading a text presupposes, or is facilitated by, some relationship to the author.

The child suggests that the text must be complemented by some knowledge of the author himself or by some magical effect induced by a close personal relationship (we know each other and are able to understand our written messages as a consequence of this friendship). The usefulness and convention of written language is overshadowed by the child's preference for direct communicative situations in which reading is unnecessary/unpractical and talking essential.

A3 The child conceptualises reading as a haphazard "bingo process".

This conception is a pseudo-technical conception of reading and implies the writing of a great number of letters or letter-like constructions that suddenly and by chance "become something" according to external judgements. This conception is very similar to a simple trial-and-error one, but with the important distinction that failure and success cannot be identified by the child.

A4 The child describes the reading process as identical to an inter- pretation of pictures

Interpretations and story telling inspired by pictures is identical to the reading act. Sometimes the child totally dis-regards the text as a vital aspect of reading.

A5 The child describes reading as identical to the reproduction from memory of a story previously read (told).

The child describes the reading act as remembering of texts (a story previously told). This conception is often linked to the interpretation of pictures where the pictures serve as concrete halting-places in a long memory-chain. There is a notion of the importance of the language form. The reproduction is not considered perfect by the child until it is in total agreement with the original version.

Those five contextual conceptions of reading describe the child's fundamental difficulty in understanding that reading depends on texts.

2.5.4.4.2 B Textual conceptions of the reading process

Conceptions previously presented in category A did not contain descriptions of reading as a text-dependent act. Category B describes children's conceptions of various textual techniques. These techniques indicate different ways of apprehending the construction of the text.

Children present two principally different ways of describing the textual construction. They focus on either the visual or the phonetic characteristics of a text. The first is B1: focus on the graphic construction of texts and the other is B2: focus on the phonetic construction of texts. Each of these two textual focuses contain three different strategies called (a) serial, (b) integrative and (c) matching strategies.

a. The child describes the integrative strategy as "a process following graphic or phonetic symbols by a step-by-step procedure in the text".

b. The child describes the integrative strategy as "a process of putting graphic or phonetic symbols together in a text".

c. The child describes the matching strategy as "a process of scanning external (text information) against internally stored letter/word pictures or sound pictures".

To sum up the category we can present the following 3 x 2 table.

	GRAPHIC FOCUS OF TEXT	PHONETIC FOCUS OF TEXT
SERIAL STRATEGY	B1 (a)	B2 (a)
INTEGRATIVE STRATEGY	B1 (b)	B2 (b)
MATCHING STRATEGY	B1 (c)	B2 (c)

If we compare those six conceptions of reading with how our writing system is constructed, we can describe them as possible but not necessarily true ways (in accordance with the alphabetical construction) of performing the reading act.

2.5.4.4.3 C *The child describes the reading process as a mental act aiming at content reflection of a text*

A shorter way of describing this conception is the expression "interactive conception of the reading process". The relation between reader and text is described as reciprocal. The text presupposes the reader and vice versa. The two main character- istics of this conception are the identification of content in texts and the active reflection upon this content.

2.5.4.4.4 D *The child describes the reading act by reference to the "reading body" that does the reading*

The child indicates that the reading process contains components of visible movements or spoken language. The child suggests the necessity to talk aloud or to move the mouth. The category also contains descriptions of how children conceptualise their development when learning to read. The internalisation of reading into a mental act begins in bodily movements which serve as a transition form to true intellectual "brain reading".

Besides categorisation of verbal statements, the pre-school study contains an observation of children's reading behaviour. Those observations were carried out when the child was reading or way trying to read different texts presented during the interview sessions. By listening to the tape recordings and using other behavioural cues, we made a categorisation in the same system of categories already presented. This new categorisation of reading behaviour did not produce any new categories in our system describing the reading process.

Describing conceptions of reading in grade 1 resulted, on the other hand, in an expansion of the presented categories of descriptions. Two new categories emerged in the analysis of the follow-up data.

2.5.4.4.5 E The child describes the analysis of text and text-segments into smaller units

We may call this conception of the reading process an "analytical reading approach". The description contains explanations of how to analyse bigger units of written language (words) into smaller ones (letters, sounds). Within the textual category previously presented, we find an integrative strategy which is detrimental to this analytical approach.

2.5.4.4.6 F The child describes the reading process by regarding reading as a prerequisite for spoken language

The child conceives of written language as a criterion for describing spoken language. Written language is the ultimate criterion to describe spoken language according to the children. The high prestige of written language influences the children's attitude to spoken language in a magical way, and written language is considered a prerequisite for spoken language.

Among the younger children (from 3 years of age) we get answers suggesting that speech is a central part of the reading process. Those children place spoken language in the same category as reading. Reading is the same as telling a story (context category). In category D (bodily reading) speech equals or precedes reading. The following statement sums up the development of the child's conceptions of the relation between spoken language and reading: pre-schoolers equal or see spoken language as a precursor to reading, while the schoolchild thinks of reading as preceding spoken language.

42% of the pre-school children give answers categorised as context-related. Those answers are reduced radically in grade 1. Textual answers in the pre-school are found in two-thirds of the interview records and in practically all interviews in grade 1. One-fifth of the pre-schoolers describe the reading process in interactive terms, while only one child makes the same description in grade 1. The main categories collapse in grade 1 and every child (except two) describes the reading process in a textual manner. The graphic integrative strategies increase from one-fifth of the answers in pre-school to two-fifths in grade 1.

In actual reading behaviour (observed during actual reading or reading trials) pre-school children demonstrate greater awareness (related to a presented definition of reading) of written language than in their verbal descriptions.

Both verbal descriptions and actual reading behaviour are uni-dimensional regarding textual focus (graphic and phonetic) in pre-school. This is not the case in grade 1, where a child can use parallel graphic and phonetic descriptions of the reading process.

2.5.5 THE ESSENCE OF READING

The empirical results of the functional and the formal aspects of reading are integrated into the essential question: "What is reading?" This essence of reading verbalised by children and integrated empirically presupposes two things. Essence refers to some kind of definition, and thereby valuation, of conceptions of reading. This criterion used on children also links conceptions of reading to adults.

When we evaluate conceptions we change from a second-order perspective to a first-order perspective. This change indicates new kinds of distinctions in the empirical material.

The criterion of the essence of reading is linked in the following ways to the above categories of describing the two reading aspects.

The functional aspect of reading: don't know answers plus category A1 (naming reading media) indicate no expression of usefulness of reading. The child may also be unable to express anything related to the reading function. In category A1 the function is not developed any further than to identifying different kinds of texts that can be read. Category A1 lacks the addendum "in order to". The qualitatively biggest category that demonstrates the child's unawareness of the reading function is category B (reading competence as a demand). The child has conceptualised the reading function, but this function is not related to any intrinsic personal use. Genuine and personal use is overshadowed by demands from people and institutions surrounding the child. Those three types of answers taken together constitute our definition of being unaware of the reading function. The other categories of the function aspect taken together constitute our definition of being aware of the reading function.

The formal aspect of reading: categories A, B and C in the formal aspect of reading are dealt with in a hierarchical manner. Context-related answers, textual and interactive answers, are different steps towards an insight into the formal aspects of reading. "Don't know" answers, together with contextual answers, are regarded as conceptions demonstrating a child's unawareness of the reading process. Descriptions of the reading process based on "things outside" the text disregard the basis of a textual system. The textual and interactive category taken together is our definition of being aware of the form of written language.

Children aware of both the functional and structural aspect of reading are defined as being "aware of the written language" (the "what" of reading). Children unaware of one or both aspects are defined as being "unaware of written language".

53 of the 61 pre-school children were interviewed and tested after one year (at the end of grade 1). The grade 1 children were placed in two groups after a strict application of the test norms of the reading tests. Stanine values less than or equal to 3 were regarded as an indication of poor reading performance. Stanine values greater than 3 were labelled good reading performance.

By this criterion we identified 28 poor and 16 good readers in grade 1. Nine of the 53 children in the follow-up study were judged as fluent or practically fluent readers in pre-school and were omitted in the analysis.

The following table shows how pre-schoolers' conceptions of reading (defined as awareness and unawareness of written language) predict their reading performance (defined as results on standardised reading tests) in grade 1.

		Observed reading performance in grade 1		
		Poor readers	Good readers	Sum
Prediction from conceptions of reading in pre-school	Predicted good readers (children aware of written language)	6	15	21
	Predicted poor readers (children unaware of written language)	22	1	23
Sum		28	16	44

In predicting the reading performance of 44 pre-school children from their awareness of the written language (function and form taken together) the prognosis is correct in 37 cases. The greatest power of the prognosis is to predict

176

poor reading skill. Poor reading perform- ance in grade 1 is predicted in 23 cases and observed by testing in 22 cases.

2.5.6 SUMMARY OF RESULTS AND DIDACTICAL CONCLUSIONS

Our results demonstrate that pre-schoolers (both readers and non-readers) conceptualise reading in many different ways. We find very few children unable to present answers to our questions about reading. Our main conclusion is that children are interested in, and think a great deal about, reading well before they have started school and acquired some reading competence. Those early conceptions (from about the age of 2-3 years) of reading represent one way of describing the first steps towards an actual reading competence. Comparing conceptions of reading internationally, we find that Swedish pre-schoolers give much the same answers as pre-schoolers one or two years younger in other countries. Such comparisons indicate that the societal organisation of where and when to start reading instruction, and the accompanying "common sense conceptions" of reading, are implemented in the child "regardless of age".

Starting school is a very important event in a child's life. It is closely knitted to learning to read and to being considered a "grown-up" child. Learning to read and learning to learn is much the same thing in the perspective of a child. The two activities are mixed, and mark the importance of how to organise reading instruction for beginners.

By using the above-presented criteria for the term AWARENESS OF WRITTEN LANGUAGE, we claim that pre-school children AWARE OF WRITTEN LANGUAGE have, on the average, greater chances to perform better when they get their first reading instruction than children UNAWARE OF WRITTEN LANGUAGE.

The main didactical question is how to facilitate children's awareness of written language. In other words, we try to answer the question of how to integrate the child's spontaneous notions of reading with the corresponding scientific concepts. We use Vygotsky's thesis about the relation between spontaneous and scientific concepts to demonstrate how this might be done. The development of the thinking processes can be described as a simultaneous and complementary process, where spontaneous concepts describe a movement from the empirical and concrete to the conscious and conditional, and the scientific concepts describing the opposite movement from awareness and conditionality to concreteness and personal experience.

The child's spontaneous concepts must be one point of departure for educational practice. The concept "zone of proximal development" describes the developmental potential of the child, and is formed by the span between the spontaneous concepts of the child and what the child can achieve successfully by maximal help from adults (Vygotsky, 1982).

A guideline in early reading instruction is to recognise that understanding the process of reading and its usefulness is relevant for the child well before the child starts school and learns how to acquire knowledge of reading. All adults must be ready to answer children's questions about reading. This passive guidance of the child should be completed with a milieu where adults take an active part in the child's education.

One major way of achieving greater awareness of written language is to use "metacognitive conversations" before starting the more structured instructions of learning to read. These conversations have to be grounded on a deep understanding of the conceptions held by children about their surrounding milieu of written language. The knowledge of children's conceptions of reading, together with a deep insight into the essence of written language, may suggest methods for the guidance of children when they start learning to read. Cognitive confusion and emotional stress might be avoided if teachers and other adults considered children's conceptions of reading as equally relevant as their own conceptions.

Our goal is to foster in the child a genuine comprehension of reading and writing as cultural activities aiming at a transmission of messages in a certain way between people in time and space. This might in turn help the child to a kind of reading (and in a broader perspective learning) focused on meaning and comprehension; a kind of reading that, to quote Vygotsky, is "relevant for life".

2.6 THE THREE STAGES OF LEARNING TO READ

by

Jacques WEISS
Institute of Educational research and Documentation
of French-speaking Switzerland
(Neuchatel, Switzerland)

2.6.1 SUMMARY

Learning to read is a complex cognitive activity which lasts a lifetime, from earliest infancy to adulthood. From being able to read only a few words at the age of two, the reader eventually acquires, as an adult, "swing-wing", that is to say versatile and varied, reading skills. A permanent feature of this complex activity is the active search for meaning in which context and textual linguistic signs are taken into account.

In the first stage of learning to read the child gradually discovers, in the course of his exchanges with readers, and his parents in particular, that there are shapes which directly refer to meanings (symbols, words), and graphic units (letters and letter groups) which have no meaning at all. In the second stage the child develops a mechanism — deciphering — which helps him to discover independently the meaning of messages, so that he is able, not to read, but to infer the most probable meaning. In the third stage the ability to read independently is assimilated and diversified according to the type of text and the reader's intentions.

2.6.2 PREAMBLE

2.6.2.1 The duality of writing

The purpose of this preamble is to define what is meant by the written word, the writing that for each of us, reader as well as non-reader, is a feature of our daily lives, which conveys a message, and to which each one of us attributes meaning. Writing includes strip cartoons, advertising (brochures and the products themselves), posters, correspondence, all kinds of signs, televised texts (TV and computer), and individual letters such as P for parking, L for learner driver, M for a chain store, etc. Writing thus consists not only of texts, sentences and words, but also of letters which, as in P for parking, for example, have a meaning of their own. In this case writing thus consists of units of meaning which are as infinite in number as words. These can be described as units which fall within the first articulation of language. These letter-symbols which belong to the realm of writing are, in addition, letters pure and simple, constituents of a code made up of a finite number of units, meaningless in themselves, but which once assembled produce the infinite number of words of a language. In this case, we can say that these units fall within the second articulation of language. There is thus a certain duality in writing and, in his search for meaning, the learner-reader must discover both aspects simultaneously.

2.6.2.2 A developmental conception of learning to read

Learning to read is a lifelong process. A child only has to be about 20 months old before he can read, that is to say, before he can attribute meaning to writing by, for example, identifying the word corresponding to a particular object; when slightly older, he can even read short sentences. G Doman (1978) and R Cohen (1985) have demonstrated the extent of very young children's ability.

Learning to read is a process which lends itself to a developmental and structuralist approach. It is characterised by the learner-reader's constant, active search for meaning. The child adopts several approaches and uses various means in learning to read, as if he were trying to solve a complex problem (trial and error, hypotheses). Meaning may be inferred from graphic signs, linguistic and semantic contexts, and from the reader's experience and perception of the text. This continuous learning process goes through several stages; from the acquisition of isolated skills to an increasingly versatile command of reading. The progression takes many years, and is clearly not confined, as is generally thought, to a single, special period in an individual's life, i.e. the first year of compulsory schooling (6-7 years).

The idea that the child learns to read at the age of 6-7 stems from a view of reading as essentially the ability to apply a precise deciphering technique, rather than to understand a message. This view is probably not without importance in determining what should be taught in the first years of compulsory education. Consequently, schools have encouraged the optional pre-school groups to proscribe any activity liable to anticipate the learning that rightfully belongs to the compulsory school. This attitude explains why a number of teachers and administrators refuse even to contemplate teaching reading skills before the start of compulsory schooling. And yet learning can start, and *does* start, well before this magic age of 6-7 years!

According to the structuralist and developmental approach, the ability to read is only one aspect of the more general ability to communicate. What in fact happens is that the child constructs its language according to its communication needs; communication is essentially oral in the first few years of life though, even in infancy, the child takes an interest in writing. From an early age (3-4 years)

180

writing can in fact be an important feature in oral exchanges with those around him, notably in the books read to him and via the encouragement he receives in his attempt to "read" certain words in his environment or to recognise certain graphic signs.

2.6.3 THE FIRST STAGE IN LEARNING TO READ: ACQUIRING UNCON-NECTED INFORMATION AND SKILLS

2.6.3.1 *The discovery of units in the first articulation of language via oral exchanges with readers*

In the various oral exchanges he builds up with his parents, the infant perceives that the stories read to him exist in books which contain various patterns, pictures and texts. He also discovers that the story stays in the book, that he can go back, unaccompanied, to the drawings, the names of characters, places, etc. The book thus appears as the thread of a story of a message, heard many times.

It is only gradually that the child discovers the nature of the relationship between that which he hears and that which is written. The research of Ferreiro and Teberosky (1979) has illustrated the complexity of these relationships. Writing is not, initially, the representation of all that is heard. It only becomes this gradually. At first, writing apparently only represents nouns, and is conceived of not as the reproduction of a speech sound, a phoneme, but as a sort of series of pictographs or patterns representing the concept referred to: a large object might be represented by a long word, and several objects by several words, for example. A triangular relationship (written –oral – reference) gradually develops as a result of the child's oral exchanges with those around him. Such exchanges are crucial in this learning process, since the meaning of the message or the word can only be supplied orally by the parents in this first stage of learning. Such exchanges consist of questions of the type: "What is written here?" What is that...or this letter? Write: dog". It is also through dialogue with a reader that the child will seek confirmation that his first attempts at reading are correct: "Does that say chicken? Does that say Alan?" Story-telling by parents or the nursery school teacher is another example of the oral presentation of writing. It is little wonder, then, that the child at that age, and even older, regards reading as the verbalisation of writing, or writing as the transcription of speech. Oral exchanges thus play a key role in this regard since they explain to the child the meaning of the concept represented by the writing.

Through all these exchanges the child quickly learns to attach a meaning to all the shapes (words and word groups) which he recognises and memorises. These are the first elements of a "mental dictionary", as it is known in America — a sort of reserve of all the words the child has discovered the meaning(s) of, and which he is able to re-read in a fraction of a second.

2.6.3.2 *The discovery of units in the second articulation of language*

In this first stage the young child also discovers that there are graphic signs which we call letters, which do not, in general, correspond to any meaning. There are instances when these letters refer to a meaning, as in "GPO", but more often than not they correspond quite simply to a sound which well-intentioned parents, anxious to teach their children how to read and write, point out to them in their exchanges and reading. The various onomatopieic words, used so often in children's literature and cartoons to illustrate the cries of animals, the rumbling of engines and other explosions and bangs, afford many perfect opportunities for teaching such matters. A survey of pre-school children conducted by Guillaume (1977) has demonstrated that a significant number of them already knew several letters of the alphabet.

2.6.3.3 *Sharpening visual perception*

It is not easy for the adult reader, accustomed to years of reading a given alphabetical script which has become familiar and obvious, to imagine or even remember the complexity of this script when it was being learnt. To jolt the reader's memory, here are extracts of different scripts:

श्राद्धस्य च विधिं ब्रह्मन् पितृणां सर्गमेव च ।
ग्रहनक्षत्रताराणां   कालांयियपरसंस्थितिम् ॥३३॥

Can any graphic signs or identical shapes be recognised in the extracts above? The task is not an easy one, and identifying letters and differentiating between words is tricky. One can easily imagine that it would take a long time to learn how to make fine distinctions between the similar graphic signs and read these texts fluently.

The same applies for someone who is learning to read the Roman script. Whilst he may well be able to distinguish "mur" from "papillon", he would need a much more complex strategy not to confuse "chapeau" with "chateau", "livre" with "litre" ... The context often helps to remove ambiguity so that fine perceptive analysis is unnecessary. There are cases, however, when the context proves to be inadequate, as, for example, for "six" – "dix" or "poule" – "poulet". In this case words are identified by reference to differentiating markers (letters, diacritic signs, first syllables).

Research on visual perception conducted some years ago by Vurpillot (1972) reveals that children only learn to distinguish different or identical shapes at a relatively late stage (around 6-7 years). The conclusion to be drawn from this is that reading, in this initial stage, is clearly still only approximative.

2.6.4 THE SECOND STAGE IN LEARNING TO READ: TOWARDS INDEPENDENT READING BY INTEGRATING PREVIOUSLY ACQUIRED KNOWLEDGE

Whereas in the previous stage the child's main strategy for unearthing the meaning of new messages (words, sentences, symbols) was preferentially to turn to a third party reader, in this stage he draws on his own resources which he developed in stage one: his "vocabulary", that is to say all the immediately recognisable shapes, and his knowledge of certain letters and letter groups. Reading a text from this moment on is like tackling a complex intellectual problem. The cognitive theories normally applicable to "problem solving" help to explain the intellectual processes underlying the act of reading.

2.6.4.1 Beginning the second atage: a heuristic approach

The behaviour of the learner-reader at the beginning of this second stage is explicable in terms of the heuristic procedures adopted in "solving complex intellectual problems". In such cases, the subject postulates various hypothetical solutions which he considers to be the most probable according to the way he construes the situation, and then checks to see if these hypotheses are correct (Weiss 1980). Thus, faced with a message, all readers postulate a hypothetical meaning which they regard as the most probable in view of the context and various semantic, syntactic and morphological signs furnished by the text, and then check it. The coherence or plausibility of the message thus determines whether the hypothesis postulated is correct. This probabilistic view has inspired most of the prevailing theories on reading.

The activity of the learner-reader at this stage basically consists of integrating the various sources of information he has available, in order to arrive at a probable hypothetical meaning. According to the extent of the incomplete knowledge he has acquired previously, the child will go on to adopt an appropriate heuristic strategy as the safest way of arriving at a probable meaning; using a code is thus an effective strategy right at the beginning of the learning process, when the child's vocabulary is still too small for him to recognise many words likely to help him infer some kind of probable meaning. If the child can only recognise a third of the words in the text his chances of understanding the message are, in fact, the same as if he had known two thirds. At the beginning of this second stage the clues to meaning are still only few and far between and imperfect, the context is still too obscure, reading too slow, and cultural references too limited for the child to be able to make any hypotheses which are in the least probable. Margins of error are still wide and areas of uncertainty too great and, as a result, extensive use is made of deciphering techniques to guide inferences and discover new meanings. These techniques are used less and less as the child's vocabulary increases, and he can consequently read faster and get a better gist of the message.

This heuristic approach can be fully appreciated by putting ourselves in the shoes of the learner-reader. That can be done by reading various texts in which only a few words and a few letters are familiar. Like the child, we thus have only part of the information and must resort to trial and error if we are to succeed in reading the message:

Texte 1

Ɣa roussice d'un renouveʋʋemenϕ podagogique pourraiϕ donc bien ėϽre fonccion du degro de cohorence des principes de ψa formaϕion avec ceux de ψa nouveψψe moϕhodoψogie.

Codes: ė = o   t = ϕ   1 = ψ

Texte 2

Au cours du recycψλge, en effeϕ, ψ'enseignλnϕ ϖeuϽ se ϖonoϽrer des λϽϽiϽudes, des comϖorϕemenϽs eϕ des modes de fλire λdoϖϽos ϖλr ψes formλϕeurs à son endroiϕ, eϕ ψes λssimiʋer ϖλr imϖrognλϕion.

Codes: ė = o   t = ϕ   1 = ψ   p = ϖ   a = λ

Texte 3

Iψ s'λgiϕ d'une imϖrognλϕiω λcϕive, dɣs ψλ mesure où ψ'ςseignɣϕ meϕ ς oeuvre ψui-même ψes ϖrinciϖes moϕhodoψogiques ϖroconisos ϖλr ψe renθveψψemςϕ ϖodλgogique.

Codes: ė = o   t = ϕ   1 = ψ   p = ϖ   a = λ   on = ω

an = ɣ   un = δ   en = ς   ou = θ

We will each have noticed that the more unfamiliar words or letters there are in a text (text 3), the more we need to resort to deciphering.

This heuristic, trial and error approach, which consists in inferring the most probable meanings and then checking them, is comparable to the general cognitive strategy which each individual uses to solve a complex problem. Moreover, it is not exclusive to writing, since it is also used for understanding oral messages (Lentin, 1978). Actually, in oral communication meaning is constructed on the basis of generally incomplete information, unfamiliar words, and words or passages not heard because of background noise or interference. In both cases the meaning must be deduced from the situation and the context, and will either be confirmed or refuted by subsequent information (missing words). The redundancy of messages facilitates the understanding of speech, whereas with writing the reader can go back to seek the missing or misinterpreted information.

2.6.4.2 Deciphering as a way of reducing uncertainty

At the beginning of this second stage, deciphering, i.e. the verbalisation of writing, is a major strategy in the search for meaning, because some words cannot be deduced from the linguistic context, which is still too uninformative.

Deciphering enables the child to discover independently the meaning of certain phonetic words in French, such as "mur, piano, Eric, Marc", as well as various onomatopoeic words or sounds (phonemes) corresponding to a graphic sign. Deciphering presupposes, however, that the child is able to connect *"syllabic fusion",* (to coin Leroy-Boussion's phrase), the sounds corresponding to the letters or letter groups. We have shown elsewhere (Weiss, 1979) that this ability only develops at a late stage.

Phonetic words are rare in French, and systematic application of the deciphering technique is thus not sufficient to enable the child to master reading. It is only a way of getting closer to a probable meaning by verbalising a word which bears some resemblance to a possible word, given the context. Deciphering is, to a certain extent, a way of reducing uncertainty and guiding the reader's hypotheses.

2.6.4.3 Vocabulary for increasing reading speed

Once a particular critical threshold, still to be defined (e.g. two thirds of the words in a text) has been crossed, vocabulary increases reading speed in the case of increasingly long texts. The short-term memory stores information for 20-30 seconds, and thus enriches the reader's heuristic strategies by furnishing information which helps him predict the meaning of words and phrases and dispense with detailed deciphering. The perception of a few signs, such as the first syllable, is sufficient to confirm that the text has been read correctly. It even becomes possible to skip certain words or expressions, or to read words which are not there. Missing words can easily be found in an incomplete text whose content is familiar. If, however, the context is difficult, the vocabulary unfamiliar, and reading slower, the missing information will not be found.

2.6.4.4 A breakthrough with reading at the end of the second stage

The end of this second stage is marked by the acquisition of a new skill which exceeds the sum of the partial skills of stage one and the beginning of stage two. This is known in educational circles as "the emergence" of reading, or the breakthrough. This phenomenon may well bear some resemblance to what *gestalt* psychologists call "insight", that is to say a sort of sudden illumination which is revealed by the unexpected manifestation of a new behaviour pattern.

In reading, this new skill consists of being able to read and understand a text unaccompanied. This skill is not the result of mastering deciphering techniques, but of the ability to attribute a meaning by exploiting various interacting pools of knowledge: an extensive vocabulary, the possibility of using a code to decipher all or parts of the words, reference to signs (the beginning of words, differentiating letters to confirm postulated hypotheses), sufficiently fast reading speed for use to be made of the short-term memory so as to be better equipped to infer the meaning of the text. Deciphering is thus necessary at this stage, but does not entirely account for the breakthrough in independent reading.

These previously separate skills are thus suddenly integrated to produce a new skill – independent reading. A child's ability to decipher a message orally has traditionally been assessed by reading aloud, but it would be a mistake to regard this ability as indicative of the breakthrough or emergence of reading.

The second stage thus ends at the point where the child displays an ability to read a written message unaccompanied. This skill is still at the embryonic stage and must be developed. It is developed during the third stage of learning to read, in which the role of the school is just as important as in the previous stages. However, teachers often feel that they have accomplished their task once the pupil has crossed the threshold of first mastering independent reading. Under pressure from the curriculum, they tend to concentrate on other subjects. This is a serious shortcoming and is probably much more responsible for adolescent illiteracy than teaching methods *per se* on which blame is so often laid. Who would dare to claim that because a swimmer can do one width of the pool he is then able to float at sea or in a river, save a drowning man or swim fully clothed...? A skill as complex as reading needs a lot of practice to make perfect.

2.6.5 THE THIRD STAGE OF LEARNING: TOWARDS MASTERY AND FLEXIBILITY

In the third stage the "primal" skill of reading, developed by the child in the previous stages, is assimilated and diversified.

2.6.5.1 *Improving comprehension by assimilating strategies for taking-in information*

Once it has been perfectly mastered, reading, like other skills (such as driving and shorthand) becomes fully assimilated and automatised. This activity can only be mastered by practising it at regular intervals and for long periods. Given that writing plays such a small part in the child's extracurricular life, a wide range of texts needs to be incorporated into the daily school timetable to ensure that this learning process is maintained: the child must be given a chance to increase his vocabulary and reading speed, develop comprehension, and structure the information obtained in order to memorise it effectively. This is in addition to maintaining and developing the child's taste for reading and literature, and making him better equipped to appreciate stylistic and aesthetic qualities. It is thus the teacher's duty to ensure that a text read quickly is well understood, and a text read aloud is interpreted correctly; children should also be taught how to appreciate literary style by reading texts aloud ("les sanglots longs..."). If the pupil

186

is to master reading adequately, more time must be devoted to reading in class beyond the second year at primary school.

It is important to make it clear that although verbalisation in reading (whatever form it takes: reading aloud, internal verbalisation or sub-vocalisation) in some cases serves no purpose and is actually an interference, in others it is absolutely essential. It would thus be erroneous to seek to divorce verbalisation entirely from all reading behaviour. This would be to forget that most creative writers, and other authors too, are concerned about the way their writing sounds; that is what is generally known as style. It is also true that there are writers who put all their effort into making their writing melodic and rhythmical (e.g. Flaubert).

2.6.5.2 Increasing flexibility in reading or developing "swing-wing" reading

With time, the child learns how to use different reading techniques according to the information he wishes to obtain and the type of text to be read. Texts can either be explored or skimmed, read silently and carefully, glanced at, verbalised internally or read aloud.

There are also several types of text. For instance, Valiquette (1979) distinguishes between 5 types of texts, and Halliday (1973), 7. According to Valiquette, (see Appendix 1), texts are either informative, persuasive, expressive, poetic or comic, whilst Halliday classifies them rather differently: he makes a separate category of instructive writing (instructions for use), directional writing (homework note-books), communicative writing (classified advertisements), personal writing (correspondence, diaries), heuristic writing (questionnaires), imaginative or projective writing (novels, poetry, puns or play on words), and, lastly, informative writing (newspapers and magazines).

There is a correspondence between the various types of reading strategies and texts (see appended table). Thus instructions, directions for use, and recipes must be read in their entirety and exhaustively, but not necessarily quickly, if they are to be clearly udnerstood; the information in railway timetables, telephone directories or dictionaries has to be picked out. A daily newspaper can be skimmed through; a thriller may be read quickly and visualised; whilst poetry would need to be internally verbalised for its charm and melody to be appreciated.

It may also be necessary to read a text in several different ways successively. A text may be skimmed to pinpoint the passages which need then to be read carefully. That applies to scientific papers, reference works such as encyclopae-dias, administrative reports, and weekly magazines. Reading behaviour may also vary according to the reader's intention, depending on whether he wants to get the gist, learn a text by heart, or summarise it.

Efficient reading, therefore, depends on the reader's being able to adopt a flexible approach to the text. This is what is learned in the third stage. It is, from this point on, up to the school to encourage and maintain this learning process; but is it doing enough?

2.6.5.3 *The school is responsible for adolescent illiteracy*

The teachers who teach first year pupils are generally successful in getting the great majority of pupils as far as the first stage of independent reading. Unfortunately, in the second year, and particularly in subsequent years, the school tends to neglect the teaching of this complex skill and is thus responsible for the collapse of the shaky foundations of reading built in the first year of primary school. The absence of a structured reading course from the second to the final year at school thus appears to be the main reason for examination failure and illiteracy among many young people.

The school is, today, the main forum for the transmission of written culture. It is the school that can offer the opportunity for reading and writing. There are hardly any other situations or places where this can happen. In the past, these cultural activities were pursued in the home and were a highly valued aspect of family life, especially for the middle-class family. They were also, to a certain degree, important in working class milieux. With the advent of the telephone for communication, and radio and then television for information and entertainment, reading and writing became demoted in the cultural life of the family. The written word now survives as a tool for training and as an instrument of administrative communication. It is precisely in the public's dealings with the administration that the illiteracy of modern times is most clearly revealed.

It is thus essential that the school should become conscious of this new role which falls to it.

2.6.6 TOWARDS A VISUAL REPRESENTATION OF THE DEVELOPMENTAL VIEW OF READING

The developmental view of reading can be represented in the form of Figure A.

2.6.6.1 *Learning to read via exchanges with readers or from the context*

Curve 1 traces the development of learning the minimum meaningful units of the first articulation of language. From being the major reading strategy at the very beginning of the learning process, it becomes appreciably less important over the years as the child adopts more economic procedures in which he can dispense with the assistance of a "third party". This strategy only becomes fully effective, however, when the child approaches the age of 5-6, because it pre-suppopses that he is able to identify subtle differences between similar shapes. It is also implemented later whenever the full meaning of a new word is grasped, without recourse to outside help, from a context which offers many clues. Reading an incomplete text demonstrates that it is possible to read words that are not there, because the context implies them so strongly.

Figure A
The development of the two basic procedures for discovering meaning throughout the three stages of the learning process.

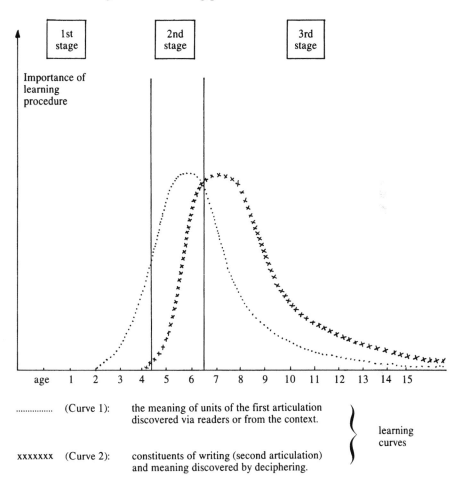

	(Curve 1):	the meaning of units of the first articulation discovered via readers or from the context.
	(Curve 2):	constituents of writing (second articulation) and meaning discovered by deciphering.

2.6.6.2 Learning to read by using a code

Curve 2 traces the process of discovering the constituents of written words
(second articulation of language), that is to say letters and letter groups. The first
meaningless units are known from an early age, but only with schooling or
parental help will the child assimilate all the graphic units (basic written forms).
However, once he can recognise a certain number of these units, he then goes on to
make his first attempts at deciphering.

These attempts are very successful when the words are written in phonetic script;
this success encourages the child to adopt this strategy when trying to understand

words which do not have this characteristic, and to produce the most probable hypotheses. Reinforced in the family and school, deciphering, which allows the learner-reader to become independent in his active search for meaning, becomes the standard reading strategy. This happens at a relatively late stage, because the child must be able to juxtapose separate phonemes, an ability which it only acquires around the age of 6-7. Deciphering remains an important strategy for grasping the meaning of words for many years, although it becomes steadily less important as vocabulary increases.

2.6.6.3 *The command of reading*

The two procedures for discovering meaning interact. Their combined effect results in the gradual development of vocabulary and improvement in reading skills. Originally separate in the first stage, the various skills acquired at that time combine to form in the second stage a new skill which is unique and original: this skill allows the child to read an unfamiliar text without help silently, or to do so fluently aloud. In the first stage this "primal" skill is assimilated, becomes automatised, and diversified into a range of skills to suit various types of texts and the many intentions of reading.

2.6.6.4 *Reference to other models*

Reading models (Giasson, Theriault, 1983), generally refer either to "bottom up" models (reading consist of translating written signs orally) or "top down" models (the written word must be learned in the context of its purpose). But there are also interactive models whereby the reader adopts different reading stragegies according to the situation and stage of his development: the bottom up process prevails when the contextual information is not forthcoming, as in reading separate words (a list), or when the child is still unable to draw on the context (first stage of learning to read and beginning of the second stage); the top down process, on the other hand, is more likely to be used for reading in context (end of the second stage and the third stage). That is certainly my view, and it is also shared by Adam (1983) and Hirsh-Pasek and Freyd (1985).

REFERENCES

Adam J-M (1983) La lecture au college. Problèmes cognitifs et textuels. In: *Enjeux.* (Namur) *3.*

Cohen R (1985) L'apprentissage précoce de la lecture: position du problème In: *Perspectives/Unesco,* XV, 1, 47-55.

Doman G (1978) *J'apprends à lire à mon bebe.* Paris: Retz.

Ferreiro E and Teberosky A (1979) *Los sistemas de escritura en el desarrollo del mino.* Mexico, Siglo XX1 (English Translation by Goodman K-J – *Literacy before schooling.* London, Exeter, NH.

Giasson J and Theriault J (1983) *Apprentissage et enseignement de la lecture* Montreal: Ville-Marie Inc.

Guillaume-Rode R (1982) *Acquisition extraordinaire de connaissances en lecture à 4-5 ans* Neuchatel, Institut romand de recherches et de documentation pédagogiques (cahier du GCR; No 3((IRDP/R 82.01).

Halliday M A K (1973) *Explorations in the functions of language.* London: Butler and Tanner

Hirsh-Pasek K and Freyd P (1985) Reading styles for deaf and hearing individuals: the importance of morphological analysis. *Literary Research Center,* Philadelphia, University of Pennsylvania, I, *1.*

Lentin L *et al* (1978) *Du parler au lire.* Paris, ESF.

Valiquette J (1979) *Les fonctions de la communication, au coeur d'une didactique renouvelée de la langue maternelle.* Quebec, Ministère de l'Education, Etudes et documents.

Vurpillot E (1973) Les facteurs perceptifs dans l'apprentissage de la lecture. *Apprendre à lire* La Tour-de-Peilz (CH): Delta.

Weiss J (1979) Quelques distinctions utiles pour un debut sur l'apprentissage de la lecture. In* *Apprentissage et pratique de la lecture à l'ecole.* Acte du Colloque de Paris, 13-14 juin 1979. Paris: Ministère de l'Education: CNDP.

Weiss J (ed) (1980). *A la recherche d'une pédagogie de la lecture.* Berne: Peter Lang.

APPENDIX

TYPES OF READING

Type of text (according to Valiquette)	Exploration of a graphic field	Pin-pointing or skimming	Integral reading	Precise verbalised reading	Rapid visualised reading
Informative					
Plan	X				
Map	X				
Guide		X	X		
Atlas					
Railway timetable		X	X		
Telephone directory		X	X		
TV viewing guide		X	X		
Dictionary		X	X		
Encylopaedia		X	X		
Recipe book			X		
Instructions			X		
Homework book			X		
Scientific papers and articles			X		
Newspapers		X			X
Administrative reports		X	X		X
Classified advertisements		X	X		X
Expressive					
Personal correspondence					X
Autobiography					X
Critical analyses					X
Poetic and literary					
Thriller				X	X
Novel				X	
Drama				X	
Poetry				X	
Novellas				X	
Stories				X	
Persuasive					
Poster	X	X			
Political or commercial propaganda texts					
Invitation			X		X
Convocation			X		

APPENDIX

TYPES OF READING

Type of text (according to Valiquette)	Exploration of a graphic field	Pin-pointing or skimming	Integral reading	Precise verbalised reading	Rapid visualised reading
Play activities					
Cartoons and riddles				x	
Puzzles				x	
Crosswords			x		

2.7 CORRELATION OF READING WITH VARIOUS FORMS OF VERBAL BEHAVIOUR

by

Dr Smiljka VASIC, Institute of Educational Research,
Belgrade, Yugoslavia

Extract from the book
"DEVELOPMENTAL SPEECH NORMS
OF OUR CHILDREN"

2.7.1 SUMMARY

Speech, language and their development represent a permanent source of interest for people in various scientific fields. The higher the quality of knowledge and the better the technique of research, the greater is the interest for this form of human behaviour which makes man what he is — "homosimbolicum" — which has remained rather a secret in both philogenetic and onthogenetic aspects of speech and language, although there is a vast number of known facts.

If the researcher's attention is focused on onthogenetic problems of speech and if it is paid to verbal development in the primary school, then there emerges the question — how much the basic forms of verbal behaviour (for example, articulation, accent, retelling, describing etc) affect reading and writing and what is the nature of relation with these higher forms of language behaviour (reading and writing)?

In our previous studies we have emphasised that the more complex forms of verbal behaviour develop together with cognitive functions in primary school and that various forms of verbal behaviour develop at their own rhythm. This development is parallel to other aspects of the child's growth and it is sometimes hardly separable. It is moreover in a certain relation with these aspects of the child's growth and it passes through all the developmental stages as they do.

We shall show in the study the following: how various forms of verbal development affect the acquisition of elementary reading as a higher form of language behaviour, the relationship between normal verbal development and the acquisition of some other forms of language development such as reading, the stages of learning to read; the acquisition speed of this form of verbal behaviour; differences which exist in reading between sexes - in more developed and undeveloped environments (town and village).

Our endeavour is to contribute in discovering relations among various forms of verbal and language behaviour; in fact, this endeavour to solve the flow and find roots of this very complex process has been found in all recent language research. Our aim, among others, is to know some of numerous basic functions of speech because it is through speech and its functions that we may approach the other forms of human behaviour (e.g. cognitive processes and emotions).

2.7.2 INTRODUCTION

According to Gray's definition[1] we know that reading is a skill which the child must be taught: firstly, to recognise the graphic symbol; secondly, to connect it with the meaning; and thirdly, to use the content it has read. The continuous process of learning to read is based on inherited and acquired environmental factors.

Many authors[2] consider good reading to be connected primarily with intelligence. Others[3] have tried to relate success in reading with eyesight quality and the physiological background of organism. Some[4] have emphasised the environment as a prevailing factor in mastering reading. We think that this complex function depends on speech, too, and on the whole series of predispositional and perceptual factors: they could be roughly grouped into organic, psychological and environmental factors.

2.7.3 THE RESEARCH

2.7.3.1 The examinee sample

As the study represents the continuation of the intensification of elementary teaching of the mother tongue, and as it is a part of a broader research named *Developmental speech norms at the primary school level*, we will not speak about the aims, tasks and general hypotheses of the research again. They are clearly defined in the introductory part of the study "On developmental speech norms in our children" and its ideological and study project.

---

[1] In the study by a team of authors in UN, published by Gray. The whole book is dedicated to these problems.
[2] Cigotsky, Piaget, Bruner.
[3] Wudvorth.
[4] Labov, Trevis.

We will first discuss the examinee sample in this study. It was the examination of a specific character. Reading was examined by a particular test. The basic examinee sample had to be smaller because of recording.[1] The principle of deliberate sample, which was applied in the earlier research, was used. Namely, the examinees from the village and town environment in Serbia, from so-called developed and underdeveloped areas, from standard, dialectical and mixed language environments were observed. There were 20 examinees of each age. The number of examinees was 700 according to our plan. Pupils from primary schools in Belgrade, Valjevo, Zajecar, Korenita and Rgotina were tested.

2.7.3.2 The language sample

Reading was observed by use of the articulatory test-story invented by the researcher according to certain principles. The starting point in constructing this story was that the test should contain all our sounds in the three characteristic positions; in initial, medial and final position.

2.7.3.3 The time and place of examination

All the examined pupils read the articulatory story-test loudly. Their reading was recorded. Each examinee said his/her name, family name, the place of schooling, the class, and then he/she started reading. Articulation, accentuation and picture description were recorded at the same time. Twenty examinees were randomly chosen from each class. Exceptions were made if there were not 10 girls or boys in a class.

All the data were qualitatively and quantitatively analysed. Each examinee's time of reading was measured by a stopwatch.[2]

Qualitative analysis consisted in marking the type of mistakes during each examinee's reading. All the mistakes were grouped according to their type. Qualitative analyses of the mistakes is given here. Besides, we defined the relationship between quantitative data: the speed of reading and number of mistakes, as well as the nature of mistakes. Indicators of reading skill development for each examinee, each age, different sexes and for different language environment (standard, dialectical and mixed one) were defined. General indicators of reading skill development for the entire examinees' sample in the Serbian province were obtained. Indicators of difficulty were established for each word from the test.

---

[1] Djordje Kostic, dr Smiljka Vasic, 1969, RAZCITAK UPOTREBE RAZLICITIH ZNAGENJA PRIDEVA "SLOBODAN" KAO FUNKCIJA UZRASTA (DEVELOPMENTAL USE OF DIFFE-RENT MEANINGS OF THE ADJECTIVE "FREE" A FUNCTION OF THE AGE), Institut za eksperimentalnu fonetiku, Beograd.

[2] It is, as we know, approximate or relative time because a certain number of seconds is lost when starting the stopwatch and a certain number of seconds is gained when ending the work; as the examiner (measurer of the time) was the same all the time, we suppose that the mistake in measuring was the permanent, personal researcher's mistake, that is a permanent feature of a measuring instrument.

It is worth saying that the same examination was repeated in a less urban environment than the Serbian province, that is the SAP Kosovo, and that these were the data on language development indicators in reading obtained for the Kosovo examinees. This material is the subject of a special study.

All the data obtained in the Serbian province were statistically analysed. The reading norm was defined for each age group; the analysis of correlation and trend was completed too. The data are graphically given because of easier recognition of the details and better legibility and comprehension. Graphical survey of statistical results is becoming more suitable for a layman to understand in some cases.[1] Illustrations help in comprehending relations which are the outcome of complex and statistical procedures.[2]

2.7.3.4 Hypotheses

We started this examination with certain hypotheses. As all the examinees were given the entire series of tests, we obtained for each of them accentuation, articulation, reading, describing, defining and comprehending the relation of opposition. Besides, we had the data on written expression. Our aim was firstly to observe each of these forms of language behaviour separately, and then to establish mutual relationships.

2.7.4 RESULTS

2.7.4.1 General

The survey of results will be given firstly according to the age, then according to the sex, different language environment and, finally, according to the sample as a whole. They all are shown in the tables and graphs.

In the first table there are shown, for example, certain frequencies of the words used in the Reading Test and in the written compositions of primary school pupils where the number of mistakes is given in percentage for each word as for the whole examinees' sample. Our experience has shown that frequency occurence of a certain word facilitates its articulation; its meaning is usually more familiar to pupils and it is thus easier to read; our aim was to construct such a test whose characteristics would be, among earlier defined ones, the feature of notional familiarity and ease for the youngest pupils as well. However, it was necessary for the test to contain words whose occurrence frequency is not prominent. The words should have been the indicators for the testing of the hypothesis that

---

[1] M.Djorovic, M Petrovic, 1974. GRAPHICAL PRESENTATION OF INFORMATION POSSIBILI-TIES FOR EFFECTIVE INTERPRETATION OF RESULTS IN AUTOMATIC INFORMATION ANALYSIS. Beograd.

[2] Such presentation was enabled by computing laboratory in Vinca in the Institute Boris Kidric and collaborators Dr.ing. M Petrovic and MA.ing. M Djorovic.

accuracy of reading depends on occurrence frequency. It has been shown that the word with the greatest occurrence frequency is, as a rule, least subject to mistakes in reading.

Besides the already-mentioned characteristics of the Story Test, its words were of different length. The shortest word was a one-syllable word-connector *i* (and); the longest word was a four-syllable word *zakrcili* (crammed). Apart from that, there were nouns (38%), verbs (20%), adjectives (16%), pronouns (8%), connectors (8%), prepositions (8%), adverts (2%), and numbers (2%). Error analysis has confirmed that the length of word is very significant for correct reading. It can be seen from the frequency word list, which has been arranged according to the percentage of mistakes, that there are mistakes mostly in four-syllabic words — that is in the longest words. However, the mistakes were not the same for all the words: the greatest number of mistakes was for the verbs, then for adverbs and finally for adjectives, numbers, prepositions, pronouns, nouns and connectors. It seems that there is the least number of mistakes in the system of nouns and then in the system of verbs. Least number of mistakes is typical for those words which were firstly acquired and which appear first in the child's language system — they are nouns and connectors. In our language, language structure nouns are, at the same time, the most frequent words.

Further analysis of the results and discussion will be followed by appropriate tables and diagrams.

2.7.4.2 Reading time diagrams

There are diagrams of reading time according to the age in the pictures 1 to 7. Comparisons are given from the aspect of functional differences in reading time as for sex, environmental levels of urbanism in a broader sense (Serbia) and somewhat narrower (Kosovo), as well as for the level of environmental development within appropriate area (village-town).

By analysing the diagram it can be concluded that speed of reading is directly dependent on the age. The time of reading goes down as the age grows. The time of reading grows only between IV and V grade. This crisis in the speed of reading is mutual for all other forms of verbal behaviour which we have examined when defining the development of verbal norms, so that we must not think of it as accidental. According to our thinking, this appearance is complicated by some biological and psychological factors which are typical for this period of development. There is a general conclusion that aggravation is greater for the village area and that it mainly occurs in IV grade, whereas for the town area it occurs in V grade.

The time of reading goes down considerably during the period of initial reading (the leap between II and III grade), and this reduction becomes permanent later, and after VII grade it reaches the level of the reading time which is needed by a mature, educated reader. Judging by the necessary reading time for a certain text, the village area population in II grade is significantly worse compared with the town area: II grade pupils in the village read the Test in 82.5 sec. and the pupils from the town read the same text in 55.5 sec. The village population masters this difference very quickly, so that already in III form it seems as though there is no significant difference. However, the environmental factor permanently influences the reading time and it can be clearly seen from these diagrams. The curves referring to the village area are always above the curves referring to the town population. It is worth mentioning that this influence is more prominent in less urbanised environment (pictures 4 to 6).

There is a comparison between the reading time as for the level of environmental level of urbanity and development of certain areas of the environment. By analysing the diagram it can firstly be seen that the longest reading time is for the village examinees, from less urbanised environment (village – Kosovo); secondly, the shortest reading time is for the town area of urbanised environment; thirdly, the flow of the curve for the reading time as for the village environment which is more urbanised, and for the town area which is also less urbanised, is approximately the same, with a notion that the time which is necessary for the Reading Test is slightly longer for the village area population. It seems as though a village in the developed area and a town in the underdeveloped area have many factors in common which influence the quality of reading in the same way; fourthly, there are seemingly no significant differences as to the reading time in III class except for the village in the most underdeveloped environment. It can be seen from the diagram that the children from this environment reach the speed of reading only in VI class (which is acquired by the III class pupils in other examined areas). The pupils from town read fastest at finishing age. Besides, the children from the village, more urbanised area, read faster than the children from less urbanised village areas. It is worth mentioning that there are no significant differences among sexes as to the speed of reading in all the observed areas.

2.7.4.3 Surveys of distributing mistakes

In the tables 1 to 4 there is the survey of the mistakes which are characteristic for certain words in the Reading Test for more urbanised environment, the village and town area as well as for the sample as a whole. There are the percentages of mistakes in Table 1 with the appropriate rank for certain words, arranged according to the order of occurrence in the Reading Test for both – the town and village area. There are mistakes for certain words in Table 2 arranged according to the rank of mistakes for the town area. The same analysis is obtained for the village area in Table 3 and there is the survey for the whole sample in Table 4. Comparing Tables 2 and 3, it can be seen that out of the first 15 words, for which there was the greatest number of mistakes, 11 words were mutual to the village and the town, although with a changed rank. The other 4 words for which there were more mistakes among the town examinees are: SADA (now), ALONG DUZ (along), IZ (from) and DJACI (pupils); among the village examinees these words were: TERET (baggage), SIROKU (wide), DVEMA (both), CESTU (road). Children from town made mistakes in words which indicate negligence in reading and may be the result of the speed of reading (for example prepositions DUZ (along) and IZ (from), then in the words which contain critical sound for the town area, for example DJACI (pupils). Children from the town do not discriminate well between the sounds *dj* and *dz*. It is the feature of their articulation, and that mistake is felt in reading as well. The village children make more mistakes in the words which are less familiar to them by their meaning and form and which are outside their experience. On the whole, it can be concluded that there are some mutual factors for both areas, such as the length of a word, the frequency usage, the sound context, unusual morphological and syntactical structures — but there are some special factors for the observed areas. Changes in word ranks which have been mistaken in both areas indicate these differences. The influence of environmental factors has been reflected in the intensity of mistakes in certain words. Deeper analysis of the intensity of mistakes for certain words in the function of age is given in the pictures 8 to 11. The pictures 8 and 9 show diagrams of mistakes for all the observed ages of the town and village area, whereas the pictures 10 and 11 show the intensity of mistakes for 15 words with most mistakes in the village and the town. It is characteristic that there are the most intensive mistakes at all ages for the same words; mistakes are more intensively made at younger ages and in the village area. It means that differences among the village and town area in reading are not prominent as for the reading time only, but as for intensity of making mistakes, too, that is, as for general quality of reading.

In pictures 10 and 11 there are spacial diagrams of mistake intensity for certain words in the function of age for the town and village area. On these diagrams there are the average times for certain ages because of comparison. Aside from giving the whole insight into the trend of intensity mistakes for each individual word in the function of age, these diagrams clearly show an emerging crisis between IV and V class, which is similar to the one for greater reading time at the mentioned ages. The intensity of mistakes at these ages is mainly greater for all individual words, even for those in which there is otherwise a small number of mistakes. By simultaneous observation of the reading time and the intensity of mistakes, there may be perceived their mutual correlation, especially at the time of so-called crisis in reading. It is interesting to note that this microanalysis of the intensity mistakes trend according to some words may indicate some psychological factors, influence on the quality of reading (such as, for example, motivation). It can be best seen when taking the word BEOGRAD. By analysing the trend of the intensity of mistakes for this word it is observed that the village children make significantly more mistakes when reading this word in the initial period (II and III class) than the children from the town. However, once they acquire it experientially (IV class), the children from the village are especially motivated to read it correctly because of connotative and affective meaning which a big city has for them: hence, they make obviously fewer mistakes in reading this word at more mature ages than the children from the town do.

The picture 8 contains the comparison of the intensity of mistakes for certain words in the town and village area. This diagram shows, too, that the children from the village fundamentally make more intensive mistakes than the children from the town. Differences are especially noticeable for words KOFERE (cases), SLEDJENI (frozen), CESTU (road), TRAMVAJI (trams), CRVEN (red), LEDE (freeze), NOSCI (porters). If we consider the nature of the mistakes, we shall see that they are at the semantic and syntactical levels. At the semantic level the mistakes are such as: KOFERE, TERET, CESTA, TRAMVAJI, NOSACI. At the syntactic level they are of the following type: SLEDJENI, CRVEN, LEDE, DVEMA. So, the children from the village make more mistakes in the Reading Test when more complex and developed language structure occur.

The town population makes more intensive mistakes for words DADA, IZ, JE, DJACI. This again leads to the former conclusion which has been made when analysing the Tables 2 and 1. Differences among the children from the town and village in the quality of reading do not reflect at higher ranks of certain words occurrences, but in the intensity of mistakes in absolute total.

The picture 9 shows that there have not been significant differences among boys and girls as to the speed and quality of reading. Somewhat greater differences have been perceived only in individual words: SADA (now), OBAVIJA (covers), LEDE (freeze), CUCORE (chat), ZAKRCILI (crammed) and NOSACI (porters) — for boys. Boys tend to shorten words (for example, SADA — SAD, OBAVIJI — OBVIJA, which shows the boys' natural general tendency of speech (faster speech). Girls make more mistakes in the following words: KONJI (horses), DJACI (pupils), LJUDI (people), JE (is), NA (on, and, I) — which can be, on the one side, interpreted to occur because of less meaningful closeness and motivational value of these words for girls (KONJI — horses) and careless reading (mistakes in reading I, NA, JE) — the mistakes which are usually not made — on the other side.

There is mutual comparison intensity for adult readers of two groups — mothers and announcers. It can be seen from this diagram that mothers intensively make mistakes in all the words and more intensively in those words in which children most frequently make mistakes: SLEDJENI(frozen), CUCORE (chat), LEDE (freeze), CRVEN (red), DVEMA (both), OBAVIJA (covers), ZAKRCILI (crammed). Announcers make few mistakes as trained readers, and if they make a mistake, the mistake occurs in the context of the words for which there are most mistakes, and that happens before or after the critical word. This can be explained by specially aiming attention of trained readers to socalled difficult words, so that there are no mistakes with these words, but there are errors before or after carefully read, difficult words. So it means that the mistake occurs before or after the word which is carefully read.

We have come to the conclusion, by more detailed statistical analysis of mistakes distribution for certain words in the Reading Test regarding the age and different areas, that this statistical process may be grouped in the category of the processes which are mathematically described by the help of distribution probability — by Ghaus' distribution. By counting characteristic statistical parameters (average value, standard deviations and others) for the town and village area and for all the ages, we have come to the conclusions which are shown on the pictures 10 and 11. There is the average time of reading for all the observed ages given on these diagrams for comparison. The analysis shows that there are intensive mistakes for all the words (II class) and that the intensity is significantly greater for the village area. Diffusion of mistakes (resolution) reduces much more quickly in the village area than in the town area although the level of making mistakes is still above the level of errors in the town area. In V class it is above the level of making mistakes in the town area. In VI class there is significant increase of intensity and the spectrum of mistakes for the town environment — resolution of mistakes is spreading — and it is directly connected with exceeding time of reading. After this crisis there is further advancement in reading, namely the resolution of mistakes narrows and the intensity lessens. After VII class statistical features of the curve for mistakes distribution become permanent so that we can consider that the process of advancing in elementary reading is completed, that is, reading at these ages produces characteristics of mature reading. By similar analysis of the diagram for the village area (picture 11) it can be seen that there is no significant aggravation of the statistical features curve in distribution of mistakes in the period of crisis (IV class) — as it happens with the town area examinees. The crisis is reflected more in the exceeding time of reading. This indicates that the mentioned crisis in the village appears in slight form and that children master it more easily. More speculatively said, it may be concluded that puberty crises are more easily mastered in the village than in the town, since the village environment is more homogeneous, and as such it postulates less demands. However, these statements should be checked with further investigations. Statistical characteristics of the reading quality are improved in VI class; so we can conclude that the process of elementary reading is over in this class already when the village area is in the matter. However, general statistical levels of mistakes among the children from the village at this age are higher than the town level of mistakes, which, in turn, indicates the significance of environmental factors influencing the process of mastering reading.

The curves for distribution of mistakes have been elaborated for adult examinees, from which it can be seen that mothers make very intensive mistakes for all the words, and that announcers make much less on a small assembly of words. For the purpose of further comparison there are, in the pictures, 12 curves for distribution of mistakes for examinees of different ages (II, V and VIII class), as well as for adults (mothers). It can be seen from this diagram that statistical characteristics of distributing mistakes in reading are approximately the same for mothers and V class pupils, with the notion that pupils make somewhat more intensive mistakes. The adequate curve for pupils of II class has considerably inferior statistical characteristics comparing it with the curve for mothers, whereas the corresponding curve for the pupils of VIII class indicates better quality of reading than mothers.

2.7.5 CONCLUSIONS

The main results of examining reading of one sample primary school population in SR Serbia (in which there have been observed at different ages from II to VIII grade - developed and underdeveloped environment, literate, dialectical and mixed on one side, and on the other side the speed of reading, nature and the number of mistakes in reading and correlation between reading, articulation, accent and description) show:

First, that there is close correlation among reading, accentuation and articulation: between articulation and accentuation the correlation co-efficient r = 0.66 (0.49% mutual factors): articulation and reading r = 0.794 (0.64%); reading and accentuation r = 0.659 (0.49%). It has been confirmed that there is not a clear connection among these three verbal activities and description; correlation co-efficient for articulation and description is r = 0.46; for accent and description r = 0.32; for reading and description r = 0.34. This means that good articulation and accentuation are conditions for good reading and comprehension, and that reading is not in any correlation with description.

Second, the speed of reading corresponds to the age, and the number of mistakes conversely corresponds to the age. It means that the speed of reading grows with the age up to IV class (in the village), and up to V class (in the town); then it becomes less at the mentioned ages and it starts growing again after IV-V class. As the age increases the number of mistakes decreases. Increasing numbers of mistakes appear in IV class in the village, that is in V class in the town. Resolution of mistakes is much more prominent in the town than in the village during the period of confusion. The reading time is getting much worse in the village; that is the variability of mistakes grows. On the other hand, improvement and better reading skill up to the end of primary schooling are noticeable in the towns. Confusion at the time of crisis in reading is less in the village, but the level of advancement in reading is already over in VI class. Practically, there is no improvement in reading from VI to VIII class and it seems as though advancing is ended. What has been achieved is achieved in VI class. It could be thought that children in the village have a quicker rhythm of development in reading and that they advance better. However, it is seeming advancement. The level at which advancement in reading is over is lower than the level of the children from the town although it has been more quickly realised. This data indicates the significance of environmental factors as to advancement in reading.

Third, there is a significant difference in regularity and speed of reading between the village and the town. The significance of the differences is at the level of 0.01 which clearly indicates the importance of environmental factors when this form of verbal behaviour is considered.

Fourth, there is no statistically significant difference between sexes in the speed or regularity of reading. If there are any differences they are not in the nature of errors, but in their intensity and distribution to individual words and their meanings, which are closer to a certain sex (for example, KONJI (horses) - critical word for girls, and SAD-SADA (now) for boys).

Fifth, the ability of description has a great variability and has in no way correlation with articulation, accent and reading both for the village and town and for the different sexes.

Sixth, in the initial period of elementary reading, mistakes in reading are arranged according to Ghaus's curve of probability, namely all the examinees make mistakes in all the words. Together with the age and mastery of reading, the mistakes group themselves around certain words whose nature imposes that if trained readers make mistakes, they do it with these words. Such words are either long or have difficult, unusual sound context or rare morphological and syntactic structure, or strange, unknown meaning or small frequency usage.

Seventh, reading is a complex activity and it does not depend on recognising graphic symbols (letters) only; it depends on many other factors, such as, apart from sensory and sensory-motor; experience, good verbal basis; a high level of general language development, intelligence and environment. It can be clearly seen from the qualitative analysis of mistakes, which beside others, shows that mistakes occur on phonetical, porphological, syntactic and semantic plane.

SECRETARIAT MEMORANDUM

prepared by the Directorate of Education,
Culture and Sport

1 INTRODUCTION

"Innovation in primary education" is the theme of a five-year project in which the 24 member States[1] of the Council for Cultural Co-operation[2] of the Council of Europe are at present taking part. This project is the main school activity on the CDCC's school programme and concerns the age-group 5/6 - 11/12. It started in 1983 and is due to last until 1987.

This project represents the first sustained effort undertaken by the Council of Europe in the field of primary education. It is the response to a need in member States to look at a period of compulsory education which has so far not often been examined and discussed in depth in the framework of an inter-governmental organisation, and which is at present the object of review or reform in most of Western Europe.

2 HISTORICAL BACKGROUND

This project fills a gap in the CDCC's recent work on the education of young children/young people. The first school project of the CDCC (1974-79) dealt with pre-school education and the early development of the child.

In the late 1970s, under the pressure of growing unemployment and its impact on the social integration of the young, the CDCC devoted its next school project to upper secondary education. The pre-school project was thus followed, from 1978 to 1982, by a project known under the title "Preparation for life", which concerned the 14-19 age group.

---

[1] Austria, Belgium, Cyprus, Denmark, France, Federal Republic of Germany, Finland, Greece, Holy See, Iceland, Ireland, Italy, Liechtenstein, Luxembourg, Malta, Netherlands, Norway, Portugal, San Marino, Spain, Sweden, Switzerland, Turkey, United Kingdom.

[2] The CDCC is the Committee responsible for educational and cultural policy within the Council of Europe. These results, together with a common concern at national level for a review of primary education, brought the Council of Europe to concentrate its attention for the next few years on the first period of compulsory education and thus, in a way, pick up the threads of the pre-school project.

The project recommended a broad approach to youth problems covering:

i   personal development;
ii   life in a democratic society;
iii world of work;
iv   cultural life.

It also indicated that, although much could be done at secondary level towards a solution of youth problems, such as unemployment and marginalisation, quite a number of them could be prevented or lessened through earlier identification, thus recognising the vital importance of the early years of compulsory education.

3 THE LICHTENSTEIN CONFERENCE[1]

In an effort to base the project on as wide a variety of information and opinions as possible, a consultation of non-governmental organisations concerned with education was held early in 1982, the results of which were presented at the first consultation of member States on the organisation of their primary education systems, the policies they have adopted and the problems they are faced with. This consultation took place at a conference on "Primary education in Western Europe: aims, trends and problems", held in Vaduz, Lichtenstein, from 9-12 November 1982. The conference confirmed the importance of the work about to be undertaken by the Council of Europe on primary education:

> "Primary education is of fundamental importance in the development of all children, and, hence, of all Europeans. It has to provide more than a narrow grounding in reading, writing and arithmetic, although these remain important. It has to widen children's perspectives of their immediate and wider physical and cultural environment. It has to help children to acquire and practise democratic values of tolerance, participation, responsibility and respect for the rights of others. It has to stimulate the development of knowledge, skills and attitudes to learning which will shape their future responses to the demands made upon them by the secondary school, by the work place and by the family and community. Primary education is, thus, a crucial foundation stage in what will be a process of life-long learning".

It showed a variety of approaches to implement innovations in primary education and identified a number of common concerns.

These can be summarised as an overall trend towards a better knowledge of children's development, bio-rhythms, and learning processes, in order to take better account of their specific needs.

---

[1] A conference is an *ad hoc* meeting at national level to illustrate the work of the Council of Europe on a given subject.
The managing group of the project (the Project Group) thus decided to concentrate its work on the innovation of primary education, devoting particular attention to its implementation.

208

4 1983: THE FIRST YEAR OF THE PROJECT

All member States were asked to submit a national report on primary education as background information for the Vaduz Conference and for the Council of Europe project in general. These reports, as well as the results of the conference itself, indicated that most countries have felt the need for an innovation of primary education and have consequently adopted, or at least elaborated, guidelines for it. The common problem, at present, seems to be implementation of innovation, i.e. the way in which the desired changes are best to be brought about.

Such a desire for change arises mainly from a better understanding of the cultural, technical and economic changes that have recently taken place in our societies and their impact on primary education.

The project consists of:

i   general reflection — through symposia — on innovation;
ii study of a restricted number of innovatory areas, e.g. new technologies, human rights education.

It agreed that this innovation should be aimed at catering for as wide a range of children's needs as possible, whether they are intrinsic to the child and his development or have arisen from recent changes in society.

A Symposium[1] was held on "Innovation of primary education in Western Europe", Han (Belgium), 17-21 October 1983, which pointed to a number of key factors for innovation processes.

A key role in the implementation of innovation is obviously played by teachers. Special attention will have to be devoted to the role they play in the innovation process, and to the type of training, both initial and in-service, that they should receive.

Of particular importance for a primary school aware of children's needs and open to its surroundings, is the kind of specialised help and advice teachers require in order to develop this new pedagogical approach. Especially crucial is the way in which this help is to be offered and structured so that full use is made of it in the classroom without infringing on the professional autonomy of the teacher.

Perhaps the most sensitive and essential type of contacts between the school and its surrounding community are the contacts with the parents. A study on the subject has been commissioned by the Project Group based on the results of several meetings, and the issue will probably be looked at in greater depth at a later stage in the project.

---

[1] A Symposium is a meeting at national level aiming at developing a number of ideas and suggestions on an important aspect of a project for further work.

In order to give the project an important practical aspect, the Project Group called for the setting-up of a school contact plan to bring into contact about ten schools undergoing comparable types of innovation in different member States. The Plan was started in 1984 and should enable the project to take into account the results of these innovation experiences and to feed into the project the reactions of teachers, educators and parents to the ideas and proposals resulting from its more theoretical and conceptual work.

It was agreed that further information on innovation experiences at present undertaken in member States should be provided in the form of case-studies, short evaluation reports of significant but perhaps isolated attempts to innovate primary education.

5 1984: THE SECOND YEAR OF THE PROJECT

1983 was the first year of the project and it can thus be defined as a year of reflection devoted to the task of translating into an operational programme the guidelines and wishes of national authorities. 1984 represented the beginning of the operational phase of the project.

The general reflection on innovation policies was continued through a symposium on "Management of innovation in primary education", Noordwijk, Netherlands, 8-12 October 1984. Starting from a close examination of the recent restructuring of primary education in the Netherlands, the need for a well-planned and explicit overall implementation policy at different levels (local, regional, national) was stressed for any successful large- scale innovation process.

A number of key factors in the innovation process were identified:

i   A well-planned system for monitoring and evaluating the innovation;
ii   a coherent teacher-training policy resulting from efficient and harmonious links between teacher-training establishments and schools' advisory services;
iii the transferability of the experience of experimental schools;
iv   continuity between primary and secondary school policies.

A series of educational research workshops on specific curriculum areas also began in 1984 with an examination of mathematics teaching (Puidoux-Chexbres, Switzerland, 1-4 May 1984), followed by a reflection of science and new technologies in the primary school (Edinburgh, 3-6 September 1984). Both research workshops clearly indicated the need for the results of education research to be brought to the knowledge of educators and policy makers in the initial phases of the innovation process and be integrated in its planning.

In 1984 the Contact School Plan came into being with the aim of:

i   exchanging innovation projects in primary education already underway in Europe;
ii obtaining the views of teachers and educators actively involved in innovation on the ideas developed within the project at the theoretical level;

The twelve schools taking part in the project are involved in projects of considerable importance at national level.

The first four case-studies were commissioned, two of them on the issue of specific needs of children, and two on the introduction of new technologies in the primary school. These are in fact two of the major concerns of the project, and the case-studies will allow the gathering of more information on the practical problems of implementing innovation in these areas. (Appendix I).

6 1985: HALF-WAY THROUGH THE PROJECT

It can be said that in 1985 the project reached "the age of maturity": all activities grew and reached the point of maximum expansion while reflection on the way in which results should be gathered and presented began.

The more theoretical activities were devoted to the issue of children's needs. A research workshop was held on "Child development" in Madrid from 17-21 September 1985, followed in October 1985 by a Symposium in Rome on "Children with special needs: the handicapped, children with learning difficulties, gifted children". While the former examined children's specific needs in the light of our more recent knowledge about process and pattern of child development, the latter looked at the ways in which children's special needs are and could be considered by member States in their educational policy.

The reflection on innovation was continued within the framework of the Contact School Plan to discuss the problems faced by a school directly involved in an innovation process, both internally and in relation to its environment.

The total number of case-studies commissioned increased to 16, and a number of additional studies were published, notably on school-family relationships.

7 SUMMARY OF THE FIRST HALF AND AN OUTLOOK ON THE SECOND HALF OF 1986

The main event of the first half of 1986 has been the Symposium on "New technologies and the primary school", held in Lyngby (Denmark) from 11-14 March 1986 which examined the impact of new technologies on society and on the school, and in particular the issue of appropriate software.

The discussion on the presentation of the results in a final report, which had begun during the third meeting of the Project Group (Lisbon, 11-14 November 1985) was continued and deepened by the Drafting Group at its first meeting (3-4 February 1986). The Group agreed that the report should first of all describe the evolution of the role of primary education, of the expectations of society from it, and its future perspective; moving then towards an explanation of the external changes which have come to bear pressure on the educational system, forcing it to try to innovate in order to adapt to its new environment and its new situation. This should be followed by the core of the report on innovation, resistance to it within the school, its environment, links within the educational system between its various levels, interaction among the actors and factors of innovation, and the resulting styles, processes and methods of innovation. A short evaluation of the project should then be introduced, followed by draft conclusions and recommendations.

The discussion will be continued within the Project Group during the current year and the next year, and the report will be presented at the Final Conference (France, October 1987).

The main events in the second half of 1986 will be the following:

i   Symposium on "Science and technology in the primary school", Cambridge (United Kingdom), 28 July - 1 August 1986:
ii Research workshop on "Reading and Writing skills", Tilburg (Netherlands), 9-12 December 1986.

8 1987: THE FINAL YEAR AND THE FINAL CONFERENCE

Apart from the Project Group, meetings will be devoted to the organisation of the Final Conference, the drafting of the final report and preliminary discussions on the dissemination of the results after the end of the project; one major meeting is envisaged for 1987: a Symposium on "The implementation of innovation in primary education at the local level", Sweden, Spring 1987.

The Final Conference of the project will take place in France during the second half of 1987. It is hoped to hold the Conference at ministerial level. The aim of the Final Conference is to inform national authorities and the educational world of the Council of Europe member States about the work and results of the project.

212

APPENDIX

LIST OF CASE STUDIES

COUNTRY	TITLE	COUNTRY OF AUTHOR
France	"The use of microcomputers in the primary school in Bar-le-Duc	Luxembourg
Greece	"The improvement of the creative and aesthetic skills of children through art and drama in the city of Athens"	Norway
Iceland	"Improvement of teaching methods and methods of evaluation	United Kingdom
United Kingdom	"The use of microcomputers in the primary schools of the city of Leeds"	Netherlands
Austria	"Spontaneous and individual learning in free choice periods (Montessori) in Salzburg"	United Kingdom
France	"Relationship between schools and between the school and its environment-ZEP (Educational priority area)	Spain
Ireland	"In-service training and additional equipment for schools in deprived areas"	Sweden
Italy	"Teaching through mass media for the acquisition of basic language skills"	Belgium
Norway	"Parent-school relationship at the primary level of the basic school in Norway"	Austria
Belgium	"School-based self-review"	Netherlands
Finland	"Delegation of decision-making from State level to local authorities"	Iceland
France	"Assistance to dispersed schools in mountain regions"	Italy
United Kingdom (Wales)	"Innovation in rural areas"	Sweden
France	"Adaptation of the pace of schooling"	Belgium
Belgium	"Time-table arrangements in French-speaking Belgium"	France
Netherlands	"Development of practical guidelines for innovation in schools"	Switzerland

LIST OF PARTICIPANTS

I   CHAIRMAN, RAPPORTEUR GENERAL AND LECTURERS

Dr. H LODEWIJKS (Chairman/President), Directeur, Foundation for Educational Research (SVO), Sweelinckplein 14, NL-2517 GK DEN HAAG.
Prof. Dr. L F W de KLERK (Rapporteur general), Katholieke Universiteit van Tilburg, Subfaculteit Psychologie, Postbus 90153, NL-5000 LE TILBURG.
Prof. Dr. E DE CORTE, Katholieke Universiteit van Leuven, Vesaliusstraat 2, B-3000 LEUVEN.
Mr Mogens JANSEN, Denmarks Paedagogiske Institut, Hermodsgade 28, DK-2200 KOEBENHAVN K.
M le Prof. Jacques FIJALKOW, Université de Toulouse-Mirail, U.E.R. des Sciences du comportement et de l'éducation, Allée A Machado, F-31058 TOULOUSE CEDEX.
Frau Prof. Dr. Mechtild DEHN, Universität Hamburg, Institut für Didaktik der Sprachen, Von-Melle-Park 8, D-2000 HAMBURG 13.
Dr. M J C MOMMERS, Institute for Educational Sciences, Catholic University of Nijmegen, Postbox 9103, NL-6500 HD NIJMEGEN.
Ms Birgita ALLARD and Mr. Bo SUNDBLAD, Enskedewagen 3, S-112 32 ENSKEDE.
Mrs. Anne SANDERSON, Language Development Centre, Sheffield City Polytechnic, 37 Clarkhouse Road, GB - SHEFFIELD S10 2LD.

II   DELEGATES

AUSTRIA
Prof. Hadmar LICHTENWALNNER, Pädagogische Akademie der Diozese St.Pölten in Krems, Dr. Gschmeidlerstrasse 22-30, A-3500 KREMS.

BELGIUM
Mme Dominique LAFONTAINE, Institut de Psychologie et des Sciences de l'éducation, Laboratoire de Pédagogie expérimentale, Université de Liège au Sart-Tilman, Bâtiment B.32 par, B-4000 LIEGE 1.

CYPRUS
Mr. Nicos LEONTIOU, Inspector of Elementary Education, Ministry of Education of the Republic of Cyprus, CY - NICOSIA.

FINLAND
Mrs. Anneli VAHAPASSI, Institute of Educational Research, University of Jyväskylä, Seminaarinkatu 15, SF-40100 JYVÄSKYLÄ.

FRANCE
M. Gérard CHAUVEAU, Responsable de recherche, Institut National de Recherche Pédagogique (INRP), 29 rue d'Ulm, F-75230 PARIS CEDEX.

FEDERAL REPUBLIC OF GERMANY
Frau Dr. Edeltraut ROEBE, University of Augsburg, Memmingerstrasse 6-14, D-8900 AUGSBURG.
Regierungsschuldirektor Hugo HERRGEN, Bezirksregierung Rheinhessen-Pfalz, Friedrich-Ebert-Strasse 14, D-6730 NEUSTADT A.D. WEIN-STRASSE.

GREECE
Excused.

IRELAND
Mr. E O'MUIRCHEARTAIGH, 47 Hillside, IRL - GREYSTONES, Co.Wicklow.

MALTA
Mr. Thomas E. FARRELL, Education Officer in the Primary School Sector of the Department of Education, Lascaris, M - VALLETTA.

NETHERLANDS
Dr G. BRENNINKMEYER, Chairman of the Board, Foundation for Educational Research (SVO), Sweelinckplein 14, NL-2517 GK DEN HAAG.
Dr. S BANDBERGEN, Ministry of Education and Science, PO Box 25000, NL-2700 LZ ZOETERMEER.
Dr. J D H VAN DONGEN, Instituut voor Sociaal-Wetenschappelijk Onderzoek, Postbus 90153, NL-5000 LE TILBURG.
H.V.D. BERG, Stichting Centrum voor Onderwijsonderzoek van de Universiteit van Amsterdam, Grote Bickerstraat 72, NL-1013 KS AMSTERDAM.
Mevr. Marjan WERTS, Stichting voor Leerplanontwikkeling (S.L.O.), Postbus 2041, NL-7500 CA ENSCHEDE.
Dr. J.H. BOONMAN, Rijksuniversiteit Utrecht, Faculteit der Sociale Wetenschappen, Postbus 80140, NL-3508 TC UTRECHT.
Dr. W VAN PEER, Department of Language and Literature, University of Tilburg, Postbus 90153, NL-5000 LE TILBURG.

PORTUGAL
M. Carlos VINCENTE, Directeur des Services de l'Enseignement primaire, Direccão-General do Ension Basico, Av. 24 de Julho 138, P - 1399 LISBOA CODEX.

SPAIN
M. Fernando ZOLLE DIAZ, Consejero técnico, Responsable del Ciclo Inicial en la Dirección General de Renovación Pedagógica, Ministerio de

Educación y Ciencia, Los Madrozos 17, E-28014 MADRID.

SWEDEN
Ms Birgita ALLARD, Enskedewagen 3, S-122 32 ENSKEDE.
Mr. Bo SUNDBLAD, Enskedewagen 3, S-122 ENSKEDE.

SWITZERLAND
M. François STOLL, Professeur de psychologie à l'Institut psychologique de l'Université de Zurich, Zürichbergstrasse 44, CH-8044 ZURICH.

UNITED KINGDOM
Prof. Michael STUBBS, University of London, Institute of Education, 20 Bedford Way, GB - LONDON WC1.

III PROJECT NO.8 OF THE COUNCIL FOR CULTURAL CO-OPERATION (CDCC)

Prof. Dr. Wilhelm WOLF, Head of Section I/la (Primary Education) Bundesministerium für Unterricht, Kunst und Sport, Minoritenplatz 5, Postfach 65, A-1014 WIEN.

IV OBSERVERS

HUNGARY
Mr. Laszlo TRENCSENYI, Hungarian National Institute of Education, Orszagos Pedagogiai Intezet, Gorkij Fasor 17-21, H-1071 BUDAPEST.

YUGOSLAVIA
Dr. Smiljka VASIC, Institute of Experimental Phonetics and Language Difficulties, Yelene Cetkovic 12, YU-11000 BELGRADE.

WORLD CONFEDERATION OF ORGANISATIONS OF THE TEACHING PROFESSION (WCOTP)
M. Jean-Bernard GICQUEL, Sécretaire Général de la FIAI, 3 rue La Rochefoucauld, F-75009 PARIS.

INTERNATIONAL FEDERATION OF TEACHERS' UNIONS
Excused.

V ORGANISERS

1 Dr. J G L C LODEWIJKS (Director) and Drs. R VERKOEYEN, Foundation for Educational Research in the Netherlands/Stichting voor Onderzoek van het Onderwijs (SVO), Sweelinckplein 14, NL-2517 GK 'S-GRAVENHAGE.

2   CATHOLIC UNIVERSITY OF TILBURG
    Prof. Dr. Len F W de KLERK, Katholieke Universiteit van Tilburg, Subfaculteit Psychologie, Postbus 90153, NL-5000 TILBURG.
    Dr SIMONS (same address).
    Mevr. I VAN DIJK (same address).

3   COUNCIL OF EUROPE
    M. Michael VORBECK, Chef de la Section de la Documentation et de la Recherche pédagogiques, Conseil de l'Europe, BP 431 R6, F-67006 STRASBOURG CEDEX.
    Mme Sylviane WEYL, Assistante (same address)
    Mlle Ellen VAN DEN HOVEN, Stagiaire (same address).

VI INTERPRETERS

    Ann MEYER, Conseil de l'Europe, BP 431 R6, F-67006 STRASBOURG CEDEX.
    Alexandrine de FRANCE, 64 Avenue Circulaire, B-1180 BRUXELLES

Printed and bound by CPI Group (UK) Ltd, Croydon, CR0 4YY

23/10/2024

01777667-0009